THE FOUR COLOUR THEOREM

ASHAY DHARWADKER

INSTITUTE OF MATHEMATICS

H-501 PALAM VIHAR, GURGAON, HARYANA 122017, INDIA

www.dharwadker.org

CONTENTS

A NEW PROOF OF THE FOUR COLOUR THEOREM

INTRODUCTION 5

I. MAP COLOURING 6

II. STEINER SYSTEMS 8

III. EILENBERG MODULES 10

IV. HALL MATCHINGS 12

V. RIEMANN SURFACES 14

VI. MAIN CONSTRUCTION 23

APPENDIX 1: COMMON SYSTEMS OF COSET REPRESENTATIVES 37

APPENDIX 2: RIEMANN SURFACES 42

APPENDIX 3: THE WITT DESIGN 43

REFERENCES 79

A NEW PROOF OF THE FOUR COLOUR THEOREM

Ashay Dharwadker

Abstract

We present a new proof of the celebrated four colour theorem using algebraic and topological methods. This proof was first announced by the Canadian Mathematical Society in 2000 and subsequently published by Orient Longman and Universities Press of India in 2008.

INTRODUCTION

The famous four colour theorem seems to have been first proposed by Möbius in 1840, later by DeMorgan and the Guthrie brothers in 1852, and again by Cayley in 1878. The problem of proving this theorem has a distinguished history, details of which abound in the literature. The statement of the theorem may be introduced as follows. In colouring a geographical map it is customary to give different colours to any two countries that have a segment of their boundaries in common. It has been found empirically that any map, no matter how many countries it contains nor how they are situated, can be so coloured by using only four different colours. The map of India requires four colours in the states bordering Madhya Pradesh.

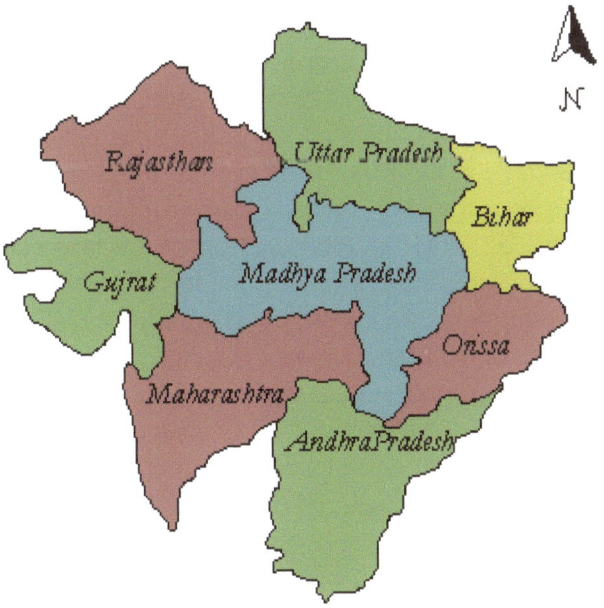

Figure 0. *Map of Madhya Pradesh and adjoining states in India*

The fact that no map was ever found whose colouring requires more than four colours suggests the mathematical theorem.

FOUR COLOUR THEOREM. *For any subdivision of the plane into non-overlapping regions, it is always possible to mark each of the regions with one of the numbers* 0, 1, 2, 3 *in such a way that no two adjacent regions receive the same number.*

STEPS OF THE PROOF: We shall outline the strategy of the new proof given in this paper. In section I on **MAP COLOURING**, we define maps on the sphere and their proper colouring. For purposes of proper colouring it is equivalent to consider maps on the plane and furthermore, only maps which have exactly three edges meeting at each vertex. Lemma 1 proves the six colour theorem using Euler's formula, showing that any map on the plane may be properly coloured by using at most six colours. We may then make the following basic definitions.

- Define N to be the minimal number of colours required to properly colour any map from the class of all maps on the plane.
- Based on the definition of N, select a specific map $\boldsymbol{m}(N)$ on the plane which requires no fewer than N colours to be properly coloured.
- Based on the definition of the map $\boldsymbol{m}(N)$, select a proper colouring of the regions of the map $\boldsymbol{m}(N)$ using the N colours 0, 1, ..., N-1.

The whole proof works with the fixed number N, the fixed map $\boldsymbol{m}(N)$ and the fixed proper colouring of the regions of the map $\boldsymbol{m}(N)$. In section II we define **STEINER SYSTEMS** and prove Tits' inequality and its consequence that if a Steiner system $S(N+1, 2N, 6N)$ exists, then N cannot exceed 4. Now the goal is to demonstrate the existence of such a Steiner system. In section III we define **EILENBERG MODULES**. The regions of the map $\boldsymbol{m}(N)$ are partitioned into disjoint, nonempty equivalence classes $\underline{0}$, $\underline{1}$, ..., $\underline{N\text{-}1}$ according to the colour they receive. This set is given the structure of the cyclic group $\boldsymbol{Z}_N = \{\underline{0}, \underline{1}, ..., \underline{N\text{-}1}\}$ under addition modulo N. We regard $\boldsymbol{Z}_N$ as an Eilenberg module for the symmetric group S_3 on three letters and consider the split extension $\boldsymbol{Z}_N]S_3$ corresponding to the trivial representation of S_3. By section IV on **HALL MATCHINGS** we are able to choose a common system of coset representatives for the left and right cosets of S_3 in the full symmetric group on $|\boldsymbol{Z}_N]S_3|$ letters. For each such common representative and for each ordered pair of elements of S_3, in section V on **RIEMANN SURFACES** we establish a certain action of the two-element cyclic group on twelve copies of the partitioned map $\boldsymbol{m}(N)$ by using the twenty-fourth root function of the sheets of the complex plane. Using this action, section VI gives the details of the **MAIN CONSTRUCTION**. The $6N$ elements of $\boldsymbol{Z}_N]S_3$ are regarded as the set of points and lemma 23 builds the blocks of $2N$ points with every set of N+1 points contained in a unique block. This constructs a Steiner system $S(N+1, 2N, 6N)$ which implies by Tits' inequality that N cannot exceed 4, completing the proof. The lemmas 1-23 and theorem 24 below are written in logical sequence. □

I. MAP COLOURING

A *map* on the sphere is a subdivision of the surface into finitely many regions. A map is regarded as *properly coloured* if each region receives a colour and no two regions having a whole segment of their boundaries in common receive the same colour. Since deformations of the regions and their boundary lines do not affect the proper colouring of a map, we shall confine ourselves to maps whose regions are bounded by simple closed polygons. For purposes of proper colouring it is equivalent to consider maps drawn on the plane. Any map on the sphere may be represented on the plane by boring a small hole through the interior of one of the regions and deforming the resulting surface until it is flat. Conversely, by a reversal of this process, any map on the plane may be represented on the sphere. Furthermore, it suffices to consider *3-regular maps*, i.e. maps with exactly three edges meeting at each vertex, by the following argument. Replace each vertex at which more than three edges meet by a small circle and join the interior of each such circle to one of the regions meeting at the vertex. A new map is obtained which is 3-regular. If this new map can be properly coloured by using at most n colours, then by shrinking the circles down to points, the desired colouring of the original map using at most n colours is obtained.

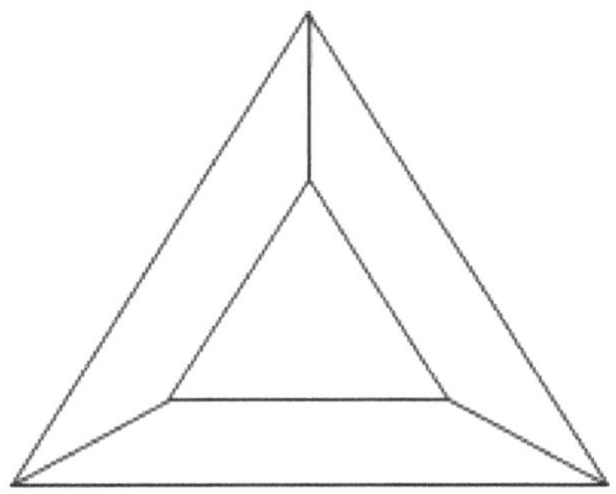

Figure 1. *A map that requires four colours to be properly coloured*

1. LEMMA. *Any map on the sphere can be properly coloured by using at most six colours.*

PROOF: Assume that the given map is 3-regular. First show that there must be at least one region whose boundary is a polygon with fewer than six sides, as follows. Let E be the number of edges, V the number of vertices, F the number of regions and F_n the number of regions whose boundary is a polygon with n sides in the given map. Then

$$F = F_2+F_3+F_4+\ldots$$
$$2E = 3V = 2F_2+3F_3+4F_4+\ldots$$

since a region bounded by n edges has n vertices and each vertex belongs to three regions. By Euler's formula $V-E+F = 2$,

$$\Rightarrow \quad 6V-6E+6F = 12$$
$$\Rightarrow \quad 4E-6E+6F = 12$$
$$\Rightarrow \quad 6F-2E = 12$$
$$\Rightarrow \quad 6(F_2+F_3+F_4+\ldots) - (2F_2+3F_3+4F_4+\ldots) = 12$$
$$\Rightarrow \quad 4F_2+3F_3+2F_4+F_5+0+\text{NEGATIVE TERMS} = 12\,.$$

Hence, at least one of F_2, F_3, F_4, F_5 must be positive. Now, if a region R with fewer than six sides is removed from the map and the resulting map coloured with six colours inductively, there is always a colour left for R. □

By lemma 1, the minimal number of colours required to properly colour any map from the class of all maps on the sphere is a well-defined natural number. We may now make the following basic definition.

DEFINITION

- Define N to be the minimal number of colours required to properly colour any map from the class of all maps on the sphere. That is, given any map on the sphere, no more than N colours are required to properly colour it and there exists a map on the sphere which requires no fewer than N colours to be properly coloured.
- Based on the definition of N, select a specific map $\boldsymbol{m}(N)$ on the sphere which requires no fewer than N colours to be properly coloured.
- Based on the definition of the map $\boldsymbol{m}(N)$, select a proper colouring of the regions of the map $\boldsymbol{m}(N)$ using the N colours 0, 1, ..., N-1.

The natural number N, the map $\boldsymbol{m}(N)$ and the proper colouring of the regions of $\boldsymbol{m}(N)$ is fixed for all future reference. By the example shown in figure 1 and lemma 1, $4 \leq N \leq 6$. The goal is to show that $N \leq 4$.

II. STEINER SYSTEMS

A *Steiner system* $S(t, k, v)$ is a set $\boldsymbol{P}$ of *points* together with a set $\boldsymbol{B}$ of *blocks* such that

- There are v points.
- Each block consists of k points.
- Every set of t points is contained in a unique block.

Note that by definition t, k, v are nonnegative integers with $t \leq k \leq v$. Steiner systems with $v = k$ (only one block that contains all the points) or $k = t$ (every k-element subset of points is a block) are called trivial. An example of a nontrivial Steiner system is $S(5, 8, 24)$ due to Witt, whose blocks are known as Golay codewords of weight eight **[see an explicit construction cf. 7]**. The group of automorphisms of $S(5, 8, 24)$ (permutations of points which permute blocks) is the largest of the Mathieu groups, M_{24}.

2. LEMMA. [J. TITS] *If there exists a nontrivial Steiner system* $S(t, k, v)$ *then*

$$v \geq (t+1)(k-t+1).$$

PROOF: First show that there exists a set X_0 of t+1 points that is not contained in any block, as follows. Suppose that for every set X of t+1 points there is a block B_X that contains it. Then, this block B_X must be the unique block containing X, since X has more than t points. Let b denote the total number of blocks. Count in two ways the number of pairs (X, B_X) where X is a set of t+1 points and B_X is the unique block containing it. One finds

$$\binom{v}{t+1} = b\binom{k}{t+1}.$$

Count in two ways the number of pairs (Y, B_Y) where Y is a set of t points and B_Y is the unique block containing it, by definition of a Steiner system. One finds

$$\binom{v}{t} = b\binom{k}{t}.$$

Hence

$$\frac{\binom{v}{t+1}}{\binom{k}{t+1}} = \frac{\binom{v}{t}}{\binom{k}{t}} = b.$$

and it follows that $b = 1$ and $k = v$, contradicting the hypothesis that the Steiner system is nontrivial. Now choose a fixed set X_0 of t+1 points that is not contained in any block. For each set Z of t points contained in X_0 there is a unique block B_Z containing Z. Each such B_Z has k–t points not in X_0 and any point not in X_0 is contained in at most one such B_Z, since two such blocks already have t–1 points of X_0 in common. The union of the blocks B_Z contains $(t+1)+(t+1)(k-t)$ points and this number cannot exceed the total number of points v. □

Recall the definition from section I that N is the minimal number of colours required to properly colour any map from the class of all maps on the sphere and ***m***(N) is a specific map which requires all of the N colours to properly colour it. The regions of the map ***m***(N) have been properly coloured using the N colours 0, 1, ..., N-1. From the map ***m***(N) and its fixed proper colouring, we shall construct a Steiner system $S(N+1, 2N, 6N)$ by defining the points and blocks in a certain way. The next lemma shows that this construction would force $N \leq 4$.

3. LEMMA. *Referring to the definition of N in section* I, *if there exists a Steiner system* $S(N+1, 2N, 6N)$ *then* $N \leq 4$.

PROOF: Since $4 \leq N \leq 6$ by definition, the Steiner system is nontrivial if it exists. By lemma 2, $6N \geq (N+1+1)(2N-N-1+1) = (N+2)N$. Hence $6 \geq N+2$ and it follows that $4 \geq N$. □

Now, the goal is to demonstrate the existence of the Steiner system $S(N+1, 2N, 6N)$ based upon the definition of the map $\boldsymbol{m}(N)$.

III. EILENBERG MODULES

Let G be a group with identity element e and let $\boldsymbol{Z}$ denote the integers. The *integral group algebra* ($\boldsymbol{Z}G$, +, -) is a ring whose elements are formal sums

$$\sum_{g \in G} n_g g$$

with g in G and n_g in $\boldsymbol{Z}$ such that $n_g = 0$ for all but a finite number of g. Addition and multiplication in $\boldsymbol{Z}G$ are defined by

$$\sum_{g \in G} n_g g + \sum_{g \in G} m_g g = \sum_{g \in G} (n_g + m_g) g,$$

$$\sum_{g \in G} n_g g \cdot \sum_{g \in G} m_g g = \sum_{g \in G} \Big(\sum_{h \in G} n_{gh^{-1}} m_h \Big) g.$$

The element n of $\boldsymbol{Z}$ is identified with the element $n \cdot e$ of $\boldsymbol{Z}G$ and the element g of G is identified with the element $1 \cdot g$ of $\boldsymbol{Z}G$, so that $\boldsymbol{Z}$ and G are to be regarded as subsets of $\boldsymbol{Z}G$. The underlying additive abelian group ($\boldsymbol{Z}G$, +) is the direct sum of copies of the integers $\boldsymbol{Z}$ indexed by elements of G. If Q is a subgroup of G then $\boldsymbol{Z}Q$ is a subring of $\boldsymbol{Z}G$ in a natural way. For each element g of G, the right multiplication $R(g)$: $G \to G$; $x \to xg$ and the left multiplication $L(g)$: $G \to G$; $x \to gx$ are permutations of the set G. Denote the group of all permutations of the set G by $Sym(G)$. Then

$$R: G \to Sym(G); g \to R(g)$$
$$L^{-1}: G \to Sym(G)\,; g \to L^{-1}(g) = L(g^{-1})$$

are embeddings of the group G in $Sym(G)$. The images $R(G)$, $L^{-1}(G)$ are called the *Cayley right and left regular representations of* G, respectively. The subgroup of $Sym(G)$ generated by the set $R(G) \cup L^{-1}(G) = \{R(g), L(g^{-1}) | g \in G\}$ is called the *combinatorial multiplication group* $Mlt(G)$ *of* G. There is an exact sequence of groups

$$1 \to C(G) \xrightarrow{\Delta} G \times G \xrightarrow{T} Mlt(G) \to 1$$

where $T(x, y) = R(x)L(y^{-1})$ and $\Delta c = (c, c)$ for an element c of the center $C(G)$ of G. If Q is a subgroup of G then the *relative combinatorial multiplication group* $Mlt_G(Q)$ *of* Q *in* G is the subgroup of $Mlt(G)$ generated by the set $R(Q) \cup L^{-1}(Q) = \{R(q), L^{-1}(q) | q \in Q\}$. The orbits of the action of $Mlt_G(Q)$ on G are the double cosets QgQ of the subgroup Q in G. The stabilizer of the identity element e is the subgroup of $Mlt_G(Q)$ generated by the set $\{T(q) = R(q)L^{-1}(q) | q \in Q\}$. A *representation* of the group Q is usually defined as a module, i.e. an abelian group (M, +), for which there is a homomorphism

T: $Q \to Aut(M, +)$ showing how Q acts as a group of automorphisms of the module. Another approach due to Eilenberg, views a module M for the group Q as follows. The set $M{\times}Q$ equipped with the multiplication

$$(m_1, q_1)(m_2, q_2) = (m_1 + m_2T(q_1), q_1q_2)$$

becomes a group $M]Q$ known as the *split extension of M by Q*. There is an exact sequence of groups

$$1 \to M \xrightarrow{\iota} M]Q \xrightarrow{\pi} Q \to 1$$

with ι: $M \to M]Q$; $m \to (m, e)$ and π: $M]Q \to Q$; $(m, q) \to q$ split by 0: $Q \to M]Q$; $q \to (0, q)$. The group action T is recovered from the split extension $M]Q$ by $mT(q)\iota = m\iota R((0, q))L^{-1}((0, q))$ for m in M and q in Q. In this context we shall call M an *Eilenberg module for the group Q*. For example, the trivial representation for the group Q is obtained by defining T: $Q \to Aut(M, +)$; $q \to 1_M$, the identity automorphism of $(M, +)$ and the corresponding split extension is the group direct product $M{\times}Q$. The Cayley right regular representation for the group Q is obtained by defining

$$T\colon\ Q \to Aut(\boldsymbol{Z}Q, +);\ q \to \Big(\sum_{g\in Q} n_g g \to \sum_{g\in Q} n_g gR(q)\Big).$$

Here $T(q) = R(q)L^{-1}(q)$ with $L^{-1}(q)$ acting trivially on the module elements and $R(q)$ acting as the usual right multiplication. The split extension $\boldsymbol{Z}Q]Q$ has multiplication given by

$$(m_1, q_1)(m_2, q_2) = (m_1 + m_2R(q_1), q_1q_2)$$

for m_1, m_2 in $\boldsymbol{Z}Q$ and q_1, q_2 in Q.

Referring to the definition in section I, N is the minimal number of colours required to properly colour any map from the class of all maps on the sphere and $\boldsymbol{m}(N)$ is a specific map that requires all of N colours to be properly coloured. Note that $\boldsymbol{m}(N)$ has been properly coloured by using the N colours 0, 1, ..., N-1 and this proper colouring is fixed. The set of regions of $\boldsymbol{m}(N)$ is then partitioned into subsets $\underline{0}$, $\underline{1}$, ..., $\underline{N\text{-}1}$ where the subset $\underline{m}$ consists of all the regions which receive the colour m. Note that the subsets $\underline{0}$, $\underline{1}$, ..., $\underline{N\text{-}1}$ are each nonempty (since $\boldsymbol{m}(N)$ requires all of the N colours to be properly coloured) and form a partition of the set of regions of $\boldsymbol{m}(N)$ (by virtue of proper colouring). Identify the set $\{\underline{0}, \underline{1}, ..., \underline{N\text{-}1}\}$ with the underlying set of the N-element cyclic group $\boldsymbol{Z}_N$ under addition modulo N. Let S_3 denote the symmetric group on three letters, identified with the dihedral group of order six generated by ρ, σ where $|\rho| = 3$ and $|\sigma| = 2$.

4. LEMMA. *($\boldsymbol{Z}_N$, +) is an Eilenberg module for the group S_3 with the trivial homomorphism*

$$T_1\colon\ S_3 \to Aut(\mathbf{Z}_N, +);\ \alpha \to 1_{\mathbf{Z}_N}$$

where $1_{\mathbf{Z}_N}$ *denotes the identity automorphism of* $\mathbf{Z}_N$. *The corresponding split extension* $\mathbf{Z}_N]S_3$ *has multiplication given by*

$$(\underline{m}_1, \alpha_1)\cdot(\underline{m}_2, \alpha_2) = (\underline{m}_1 + \underline{m}_2, \alpha_1\alpha_2)$$

and is a group isomorphic to the direct product $\mathbf{Z}_N \times S_3$.

PROOF: Follows from definition. □

Referring to section II, the goal is to construct a Steiner system $S(N+1, 2N, 6N)$. We shall take the point set of the Steiner system to be the underlying set of the split extension $\mathbf{Z}_N]S_3$. The following lemma is used in section V.

5. LEMMA. *Let* $(\mathbf{Z}(\mathbf{Z}_N]S_3), +)$ *and* $(\mathbf{Z}S_3, +)$ *denote the underlying additive groups of the integral group algebras* $\mathbf{Z}(\mathbf{Z}_N]S_3)$ *and* $\mathbf{Z}S_3$, *respectively. Then* $(\mathbf{Z}(\mathbf{Z}_N]S_3), +)$ *is an Eilenberg module for the group* $(\mathbf{Z}S_3, +)$ *with the trivial homomorphism*

$$T_2\colon(\mathbf{Z}S_3, +) \to Aut(\mathbf{Z}(\mathbf{Z}_N]S_3), +);\ \sum_{\alpha\in S_3} n_\alpha\alpha \to 1_{\mathbf{Z}(\mathbf{Z}_N]S_3)}$$

where $1_{\mathbf{Z}(\mathbf{Z}_N]S_3)}$ *denotes the identity automorphism of* $(\mathbf{Z}(\mathbf{Z}_N]S_3), +)$. *The corresponding split extension* $\mathbf{Z}(\mathbf{Z}_N]S_3)]\mathbf{Z}S_3$ *has multiplication given by*

$$\Big(\sum_{(\underline{m},\beta)\in\mathbf{Z}_N]S_3} n_{(\underline{m},\beta)}(\underline{m},\beta),\ \sum_{\alpha\in S_3} n_\alpha\alpha\Big)\Big(\sum_{(\underline{m},\beta)\in\mathbf{Z}_N]S_3} n'_{(\underline{m},\beta)}(\underline{m},\beta),\ \sum_{\alpha\in S_3} n'_\alpha\alpha\Big)$$

$$= \Big(\sum_{(\underline{m},\beta)\in\mathbf{Z}_N]S_3} (n_{(\underline{m},\beta)}+n'_{(\underline{m},\beta)})(\underline{m},\beta),\ \sum_{\alpha\in S_3}(n_\alpha+n'_\alpha)\alpha\Big)$$

and is a group isomorphic to the direct product $(\mathbf{Z}(\mathbf{Z}_N]S_3)\times\mathbf{Z}S_3, +)$.

PROOF: Follows from definition. □

IV. HALL MATCHINGS

Let Γ be a bipartite graph with vertex set $V = X\cup Y$ and edge set E (every edge has one end in X and the other end in Y). A *matching from X to Y in* Γ is a subset M of E such that no vertex is incident with more than one edge in M. A matching M from X to Y in Γ is called *complete* if every vertex in X is incident with an edge in M. If A is a subset of V then let $adj(A)$ denote the set of all vertices adjacent to a vertex in A.

6. LEMMA. [P. HALL] *If* $|adj(A)| \geq |A|$ *for every subset* A *of* X *then there exists a complete matching from* X *to* Y *in* Γ.

PROOF: A matching from X to Y in Γ with $|M| = 1$ always exists by choosing a single edge in E. Let M be a matching from X to Y in Γ with m edges, $m < |X|$. Let $x_0 \in X$ such that x_0 is not incident with any edge in M. Since $|adj(\{x_0\})| \geq 1$, there is a vertex y_1 adjacent to x_0 by an edge in $E \backslash M$. If y_1 is not incident with an edge in M, then stop. Otherwise, let x_1 be the other end of such an edge. If $x_0, x_1, ..., x_k$ and $y_1, ..., y_k$ have been chosen, then since $|adj(\{x_0, x_1, ..., x_k\})| \geq k+1$, there is a vertex y_{k+1}, distinct from $y_1, ..., y_k$, that is adjacent to at least one vertex in $\{x_0, x_1, ..., x_k\}$. If y_{k+1} is not incident with an edge in M, then stop. Otherwise, let x_{k+1} be the other end of such an edge. This process must terminate with some vertex, say y_{k+1}. Now build a simple path from y_{k+1} to x_0 as follows. Start with y_{k+1} and the edge in $E \backslash M$ joining it to, say x_{i_1} with $i_1 < k+1$. Then add the edge in M from x_{i_1} to y_{i_1}. By construction, y_{i_1} is joined by an edge in $E \backslash M$ to some x_{i_2} with $i_2 < i_1$. Continue adding edges in this way until x_0 is reached. One obtains a path $y_{k+1}, x_{i_1}, y_{i_1}, x_{i_2}, y_{i_2}, ..., x_{i_r}, y_{i_r}, x_0$ of odd length $2r+1$ with the $r+1$ edges $\{y_{k+1}, x_{i_1}\}, \{y_{i_1}, x_{i_2}\}, ..., \{y_{i_r}, x_0\}$ in $E \backslash M$ and the r edges $\{x_{i_1}, y_{i_1}\}, ..., \{x_{i_r}, y_{i_r}\}$ in M. Define

$$M' = (M \backslash \{\{x_{i_1}, y_{i_1}\}, ..., \{x_{i_r}, y_{i_r}\}\}) \cup \{\{y_{k+1}, x_{i_1}\}, \{y_{i_1}, x_{i_2}\}, ..., \{y_{i_r}, x_0\}\}.$$

Then M' is a matching from X to Y in Γ, with $|M'| = |M| - r + r + 1 = |M| + 1$. Repeating this process a finite number of times must yield a complete matching from X to Y in Γ. □

7. LEMMA. *Referring to section* III, *let* $Sym(\boldsymbol{Z}_N]S_3)$ *denote the group of all permutations of the underlying set of the split extension* $\boldsymbol{Z}_N]S_3$ *of lemma* 4. *Then* S_3 *embeds in* $Sym(\boldsymbol{Z}_N]S_3)$ *via the Cayley right regular representation.*

PROOF: Note that $S_3 = \{(\underline{0}, \alpha) | \alpha \in S_3\}$ is a subgroup of $\boldsymbol{Z}_N]S_3$. Since S_3 embeds in $Sym(S_3)$ via the Cayley right regular representation $\alpha \rightarrow R(\alpha)$ and $Sym(S_3)$ is a subgroup of $Sym(\boldsymbol{Z}_N]S_3)$, the lemma follows. □

8. LEMMA. *By lemma* 7, *regard* S_3 *as a subgroup of* $Sym(\boldsymbol{Z}_N]S_3)$. *There exists a common system of coset representatives* $\varphi_1, ..., \varphi_k$ *such that* $\{\varphi_1 S_3, ..., \varphi_k S_3\}$ *is the family of left cosets of* S_3 *in* $Sym(\boldsymbol{Z}_N]S_3)$ *and* $\{S_3\varphi_1, ..., S_3\varphi_k\}$ *is the family of right cosets of* S_3 *in* $Sym(\boldsymbol{Z}_N]S_3)$.

PROOF: By Lagrange's theorem the left cosets of S_3 partition $Sym(\boldsymbol{Z}_N]S_3)$ into $k = [Sym(\boldsymbol{Z}_N]S_3):S_3]$ disjoint nonempty equivalence classes of size $|S_3| = 6$. The same is true of the right cosets. Define a bipartite graph Γ with vertices $X \cup Y$ where $X = \{\psi_1 S_3, ..., \psi_k S_3\}$ is the family of left cosets of S_3 in $Sym(\boldsymbol{Z}_N]S_3)$ and $Y = \{S_3\psi'_1, ..., S_3\psi'_k\}$ is the family of right cosets of S_3 in $Sym(\boldsymbol{Z}_N]S_3)$ with an edge $\{\psi_i S_3, S_3\psi'_j\}$ if and only if $\psi_i S_3$ and $S_3\psi'_j$ have nonempty intersection. Note that we can select representatives of the left cosets that belong to distinct right cosets **[see a proof of this fact cf. 8]**. For any subset $A = \{\psi_{i_1} S_3, ..., \psi_{i_r} S_3\}$ of X, one has $\psi_{i_1} \in \psi_{i_1} S_3, ..., \psi_{i_r} \in \psi_{i_r} S_3$ and there exist distinct $j_1, ..., j_r$ such that $\psi_{i_1} \in S_3\psi'_{j_1}, ..., \psi_{i_r} \in S_3\psi'_{j_r}$. Hence, in the graph Γ, $|adj(A)| \geq |A|$. Hall's hypothesis of lemma 6 is satisfied and there exists a complete matching from X to Y in Γ. This is precisely the statement that a common system of coset representatives $\varphi_1, ..., \varphi_k$ exists. □

V. RIEMANN SURFACES

Let $\boldsymbol{C}$ denote the complex plane. Consider the function $\boldsymbol{C} \rightarrow \boldsymbol{C}$; $z \rightarrow w = z^n$, where $n \geq 2$. There is a one-to-one correspondence between each sector

$$\{z|(k\text{-}1)2\pi/n < arg\, z < k2\pi/n\} \text{ (k = 1, ..., } n)$$

and the whole w-plane except for the positive real axis. The image of each sector is obtained by performing a cut along the positive real axis; this cut has an upper and a lower edge. Corresponding to the n sectors in the z-plane, take n identical copies of the w-plane with the cut. These will be the *sheets of the Riemann surface* and are distinguished by a label k which serves to identify the corresponding sector. For $k = 1, ..., n\text{-}1$ attach the lower edge of the sheet labeled k with the upper edge of the sheet labeled k+1. To complete the cycle, attach the lower edge of the sheet labeled n to the upper edge of the sheet labeled 1. In a physical sense, this is not possible without self-intersection but the idealized model shall be free of this discrepancy. The result of the construction is a *Riemann surface* whose points are in one-to-one correspondence with the points of the z-plane **[see a geometric model cf. 15]**. This correspondence is continuous in the following sense. When z moves in its plane, the corresponding point w is free to move on the Riemann surface. The point $w = 0$ connects all the sheets and is called the *branch point*. A curve must wind n times around the branch point before it closes. Now consider the n-valued relation

$$z = {}^n\sqrt{w}.$$

To each $w \neq 0$, there correspond n values of z. If the w-plane is replaced by the Riemann surface just constructed, then each complex number $w \neq 0$ is represented by n points of the Riemann surface at superposed positions. Let the point on the uppermost sheet represent the principal value and the other n-1 points represent the other values. Then $z = {}^n\sqrt{w}$ becomes a single-valued, continuous, one-to-one correspondence of the points of the Riemann surface with the points of the z-plane. Now recall the definition of the map $\boldsymbol{m}(N)$ from section I. The map $\boldsymbol{m}(N)$ is on the sphere. Pick a region and deform the sphere so that both 0 and ∞ are two distinct points inside this region when the sphere is regarded as the extended complex plane. Using the stereographic projection one obtains the map $\boldsymbol{m}(N)$ on the complex plane $\boldsymbol{C}$ with the region containing 0 and ∞ forming a "sea" surrounding the other regions which form an "island". Put this copy of $\boldsymbol{C}$ on each sheet of the Riemann surface corresponding to $w = z^n$. The branch point lies in the "sea". The inverse function $z = {}^n\sqrt{w}$ results in n copies of the map $\boldsymbol{m}(N)$ on the z-plane in the sectors

$$\{z|(k\text{-}1)2\pi/n < arg\, z < k2\pi/n\} \; (k = 1, ..., n).$$

The origin of the z-plane lies in the n "seas".

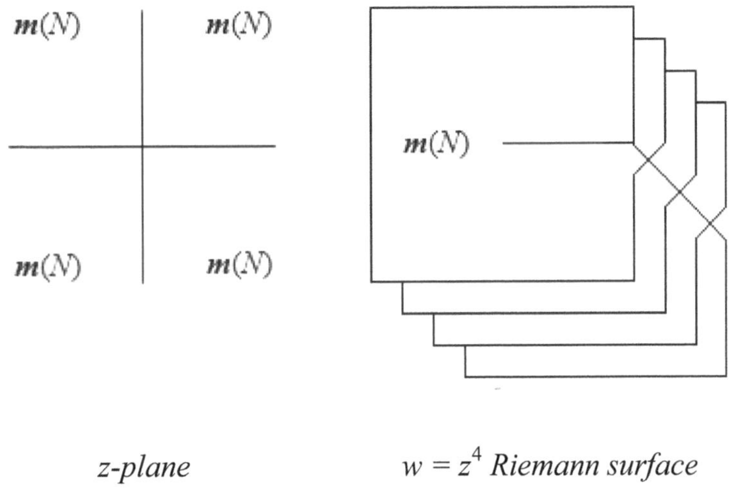

Figure 2. *An example with n* = 4

Referring to section III, the full symmetric group $Sym(\boldsymbol{Z}_N]S_3)$ acts faithfully on the set $\boldsymbol{Z}_N]S_3$. The action of an element ψ of $Sym(\boldsymbol{Z}_N]S_3)$ on an element $(\underline{m}, \alpha)$ of $\boldsymbol{Z}_N]S_3$ will be written as $(\underline{m}, \alpha)\psi$. This action extends to the integral group algebra $\boldsymbol{Z}(\boldsymbol{Z}_N]S_3)$ by linearity

$$\Big(\sum_{(\underline{m},\alpha)\in \boldsymbol{Z}_N]S_3} n_{(\underline{m},\alpha)}(\underline{m}, \alpha)\Big)\psi = \sum_{(\underline{m},\alpha)\in \boldsymbol{Z}_N]S_3} n_{(\underline{m},\alpha)}((\underline{m}, \alpha)\psi).$$

Referring to lemma 8, fix a common coset representative φ_i of S_3 in $Sym(\boldsymbol{Z}_N]S_3)$ and fix a pair $(\beta, \gamma)\in S_3\times S_3 = Mlt(S_3)$. There are two cases depending on whether $\beta = \gamma$ or whether $\beta \neq \gamma$.

CASE 1. Suppose $\beta \neq \gamma$. Consider the composition of the functions

$$\boldsymbol{C} \to \boldsymbol{C};\, z \to t = z^2 \text{ and } \boldsymbol{C} \to \boldsymbol{C};\, t \to w = t^{12}.$$

The composite is given by the assignment

$$z \to t = z^2 \to w = t^{12} = z^{24}.$$

There are twenty-four superposed copies of the map $\boldsymbol{m}(N)$ on the w-Riemann surface corresponding to the sectors

$$\{z|(k\text{-}1)2\pi/24 < arg\, z < k2\pi/24\}\ (k = 1, ..., 24)$$

on the z-plane. These are divided into two sets. The first set consists of twelve superposed copies of the map $\boldsymbol{m}(N)$ corresponding to the sectors

$$\{z|(k\text{-}1)2\pi/24 < arg\, z < k2\pi/24\}\ (k = 1, ..., 12)$$

of the upper half of the z-plane which comprise the upper sheet of the t-Riemann surface.

The second set consists of twelve superposed copies of the map $\boldsymbol{m}(N)$ corresponding to the sectors

$$\{z|(k-1)2\pi/24 < arg\ z < k2\pi/24\}\ (k = 13, ..., 24)$$

of the lower half of the z-plane which comprise the lower sheet of the t-Riemann surface.

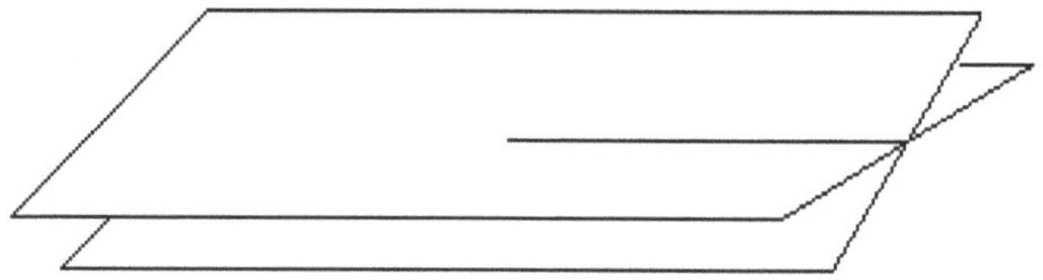

Figure 3. *Sheets of the t-Riemann surface*

Label the twelve sectors of the upper sheet of the *t*-Riemann surface by elements of $\mathbf{Z}(\mathbf{Z}_N]S_3)]\mathbf{Z}S_3$ as shown:

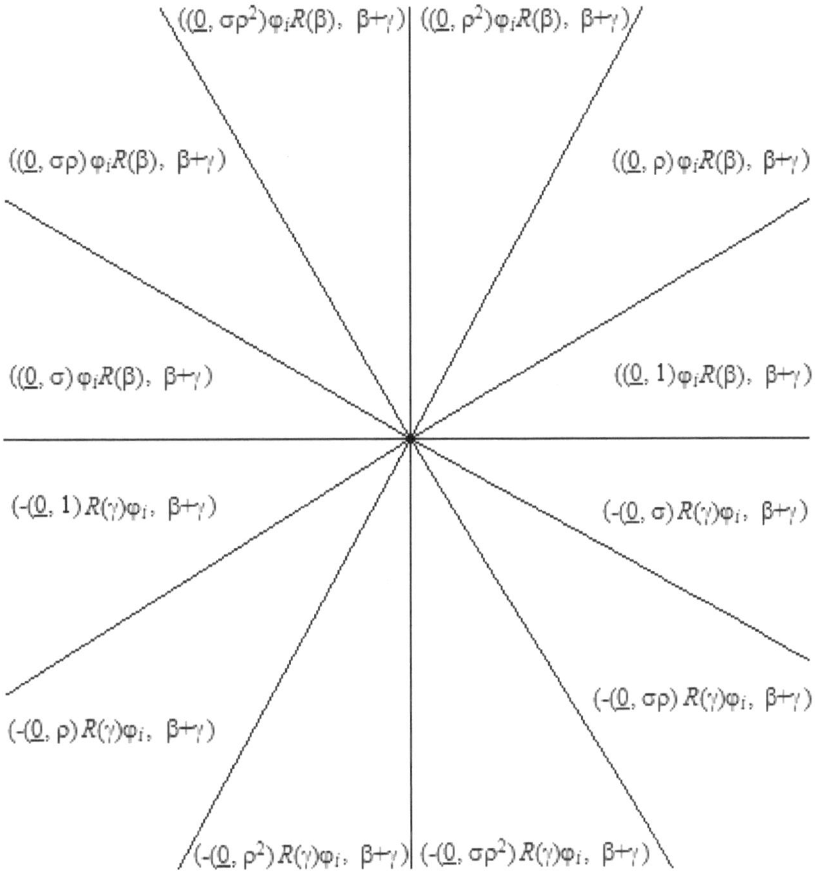

Figure 4. *Upper sheet of the t-Riemann surface*

Label the twelve sectors of the lower sheet of the t-Riemann surface by elements of $\boldsymbol{Z}(\boldsymbol{Z}_N]S_3)]\boldsymbol{Z}S_3$ as shown:

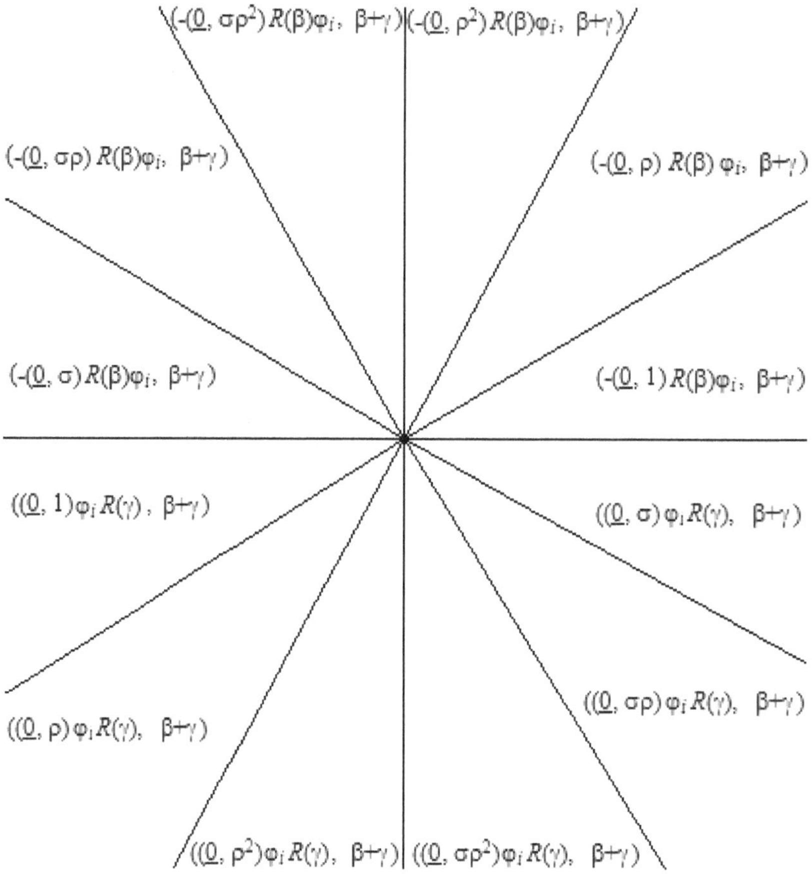

Figure 5. *Lower sheet of the t-Riemann surface*

Referring to section II, the regions of the map $\boldsymbol{m}(N)$ have been partitioned into disjoint, nonempty equivalence classes $\underline{0}$, $\underline{1}$, ..., $\underline{N\text{-}1}$ and this set of equivalence classes forms the underlying set of the cyclic group $\boldsymbol{Z}_N$. Hence there are twelve copies of $\boldsymbol{Z}_N$ on the upper sheet and twelve copies of $\boldsymbol{Z}_N$ on the lower sheet of the t-Riemann surface. The copies of $\boldsymbol{Z}_N$ are indexed by the elements of $\boldsymbol{Z}(\boldsymbol{Z}_N]S_3)]\boldsymbol{Z}S_3$ which label the sectors on a particular sheet. The branch point of the t-Riemann surface is labeled by the element $(0, \beta+\gamma)$ of $\boldsymbol{Z}(\boldsymbol{Z}_N]S_3)]\boldsymbol{Z}S_3$ where 0 denotes the zero element of $\boldsymbol{Z}(\boldsymbol{Z}_N]S_3)$.

9. LEMMA. *Referring to lemma 8, fix a common representative φ_i of the left and right cosets of S_3 in $Sym(\boldsymbol{Z}_N]S_3)$. Fix a pair $(\beta, \gamma)\in S_3\times S_3$ with $\beta\neq\gamma$. Referring to lemma 5, define a subset $T_{(\beta, \gamma)}$ of $\boldsymbol{Z}(\boldsymbol{Z}_N]S_3)]\boldsymbol{Z}S_3$ as follows:*

$$T_{(\beta,\gamma)} = \left\{((\underline{m}, \alpha), \beta+\gamma) | (\underline{m}, \alpha)\in \boldsymbol{Z}_N]S_3\right\} \cup \left\{(0, \beta+\gamma)\right\} \cup \left\{(-(\underline{m}, \alpha), \beta+\gamma) | (\underline{m}, \alpha)\in \boldsymbol{Z}_N]S_3\right\}.$$

Referring to the preceding discussion, consider the composite function

$$z \to t = z^2 \to w = t^{12} = z^{24}$$

of the complex z-plane to the w-Riemann surface. There is a copy of the set $T_{(\beta, \gamma)}$ on the upper sheet and a copy of the set $T_{(\beta, \gamma)}$ on the lower sheet of the t-Riemann surface according to the labels of the sectors in figures 4 and 5 with the branch point labelled by the element $(0, \beta+\gamma)$ of both copies. The rotation of the z-plane by π radians induces a permutation

$$p\colon\ T_{(\beta,\gamma)} \to T_{(\beta,\gamma)}$$

given by

$$\begin{aligned}(-(\underline{m}, \alpha)R(\gamma)\varphi_i, \beta+\gamma)p &= ((\underline{m}, \alpha)\varphi_i R(\gamma), \beta+\gamma)\\ (0, \beta+\gamma)p &= (0, \beta+\gamma)\\ ((\underline{m}, \alpha)\varphi_i R(\beta), \beta+\gamma)p &= (-(\underline{m}, \alpha)R(\beta)\varphi_i, \beta+\gamma)\end{aligned}$$

for all $(\underline{m}, \alpha)\in \boldsymbol{Z}_N]S_3$, such that each point of the copy of $T_{(\beta, \gamma)}$ on the upper sheet moves continuously along a circular curve that winds exactly once around the branch point, to the point superposed directly below it on the copy of $T_{(\beta, \gamma)}$ on the lower sheet of the t-Riemann surface.

PROOF: $T_{(\beta, \gamma)}$ is seen to be a well-defined subset of $\boldsymbol{Z}(\boldsymbol{Z}_N]S_3)]\boldsymbol{Z}S_3$ by setting the appropriate coefficients to zero in a typical element as described in lemma 5. Each of $R(\gamma)\varphi_i$, $\varphi_i R(\gamma)$, $\varphi_i R(\beta)$, $R(\beta)\varphi_i$ are permutations of the set $\boldsymbol{Z}_N]S_3$ and the rotation of the z-plane by π radians clearly induces a permutation p of the set $T_{(\beta, \gamma)}$ as described. □

10. LEMMA. *Referring to lemma 9, let* $Sym(T_{(\beta,\gamma)})$ *denote the full permutation group of the set* $T_{(\beta,\gamma)}$. *Let* $<p>$ *denote the cyclic subgroup of* $Sym(T_{(\beta,\gamma)})$ *generated by p. Then* $<p>$ *is nontrivial and acts faithfully on the set* $T_{(\beta,\gamma)}$.

PROOF: If $p = 1$, then $(-(\underline{0},1)R(\gamma)\varphi_i, \beta+\gamma) = (-(\underline{0},1)R(\gamma)\varphi_i, \beta+\gamma)p = ((\underline{0},1)\varphi_i R(\gamma), \beta+\gamma)$ which implies that $-(\underline{0},1)R(\gamma)\varphi_i = (0,1)\varphi_i R(\gamma)$ in $\boldsymbol{Z}(\boldsymbol{Z}_N]S_3)$. This is impossible since $1 \neq -1$ in $\boldsymbol{Z}$. Hence $p \neq 1$. Since the full permutation group $Sym(T_{(\beta,\gamma)})$ acts faithfully on $T_{(\beta,\gamma)}$, so does its subgroup $<p>$. □

11. LEMMA. *Referring to lemma 9 and lemma 10, let* 1: $\boldsymbol{C} \to \boldsymbol{C}$; $z \to z$ *denote the identity and* π: $\boldsymbol{C} \to \boldsymbol{C}$; $z \to -z$ *denote the rotation through an angle of* π *radians of the z-plane. Then the two-element cyclic group* $\{1, \pi\}$ *acts faithfully on the set* $<p>$ *as follows:* $p^n 1 = p^n$ *and* $p^n\pi = p^{1-n}$, *for all n in* $\boldsymbol{Z}$.

PROOF: The set $\{1, \pi\}$ forms a two-element cyclic group $<\pi>$ under function composition. To show that $\{1, \pi\}$ acts on $<p>$ as defined, observe that $(p^n\pi)\pi = (p^{1-n})\pi = p^{1-(1-n)} = p^{1-1+n} = p^n = p^n 1 = p^n(\pi\pi)$, for all n in $\boldsymbol{Z}$. To show that the action is faithful, let $\theta \in \{1, \pi\}$. If θ belongs to the kernel of the action then $p^n\theta = p^n$ for all n in $\boldsymbol{Z}$ so that $p\theta = p$ which implies that $\theta = 1$, since $p \neq 1$ by lemma 10. □

12. LEMMA. *Putting together lemma 9, lemma 10 and lemma 11, there is a well-defined action of the two-element cyclic group* $\{1, \pi\}$ *on the set* $T_{(\beta,\gamma)}$ *given by*

$$((\underline{m}, \alpha)\varphi_i R(\gamma), \beta+\gamma)1 = ((\underline{m}, \alpha)\varphi_i R(\gamma), \beta+\gamma)$$
$$(0, \beta+\gamma)1 = (0, \beta+\gamma)$$
$$(-(\underline{m}, \alpha)R(\beta)\varphi_i, \beta+\gamma)1 = (-(\underline{m}, \alpha)R(\beta)\varphi_i, \beta+\gamma)$$

and

$$((\underline{m}, \alpha)\varphi_i R(\gamma), \beta+\gamma)\pi = (-(\underline{m}, \alpha)R(\gamma)\varphi_i, \beta+\gamma)$$
$$(0, \beta+\gamma)\pi = (0, \beta+\gamma)$$
$$(-(\underline{m}, \alpha)R(\beta)\varphi_i, \beta+\gamma)\pi = ((\underline{m},\alpha)\varphi_i R(\beta), \beta+\gamma)$$

for all $(\underline{m}, \alpha)$ *in* $\boldsymbol{Z}_N]S_3$. *This action is faithful.*

PROOF: For each $x \in T_{(\beta,\gamma)}$, let $Orb(x) = \{xp^n | n \in \boldsymbol{Z}\}$ denote the orbit of x under $<p>$. The collection $\{Orb(x) | x \in T_{(\beta,\gamma)}\}$ forms a partition of the set $T_{(\beta,\gamma)}$ as follows. Let $x, y \in T_{(\beta,\gamma)}$. If $z \in Orb(x) \cap Orb(y)$ then $z = xp^n = yp^m$ for some $m, n \in \boldsymbol{Z}$. This implies $xp^{n-m} = y \Rightarrow y \in Orb(x) \Rightarrow Orb(y) \subseteq Orb(x)$ and $yp^{m-n} = x \Rightarrow x \in Orb(y) \Rightarrow Orb(x) \subseteq Orb(y)$. Hence $Orb(x) = Orb(y)$. Also each $x \in T_{(\beta,\gamma)}$ belongs to an orbit, namely $x \in Orb(x)$. Hence the orbits are disjoint, nonempty and their union is all of the set $T_{(\beta,\gamma)}$. For each fixed $x \in T_{(\beta,\gamma)}$ define

$$\pi: Orb(x) \to Orb(x); (xp^n)\pi = xp^{1-n}.$$

Then π is well-defined, since $xp^n = xp^m \Rightarrow xp^{n-m} = x \Rightarrow xp^{-m} = xp^{-n} \Rightarrow xp^{1-m} = xp^{1-n} \Rightarrow (xp^n)\pi = (xp^m)\pi$. Also, π is a permutation of $Orb(x)$ with $\pi^2 = 1$, since for each

$xp^n \in Orb(x)$ we have $((xp^n)\pi)\pi = (xp^{1-n})\pi = xp^{1-(1-n)} = xp^n$. Now define

$$\pi\colon\ T_{(\beta,\gamma)} \rightarrow T_{(\beta,\gamma)};\ (xp^n)\pi = xp^{1-n}$$

orbit by orbit. Then, since $\{Orb(x) | x \in T_{(\beta,\gamma)}\}$ forms a partition of $T_{(\beta,\gamma)}$, π is a well-defined permutation of $T_{(\beta,\gamma)}$ with $\pi^2 = 1$, the identity permutation of $T_{(\beta,\gamma)}$. Hence, using the definition of p in lemma 9, we obtain an action of the two-element cyclic group $\{1, \pi\}$ on $T_{(\beta,\gamma)}$ as follows. For all $(\underline{m}, \alpha)$ in $\boldsymbol{Z}_N]S_3$ define

$$\begin{aligned} ((-(\underline{m}, \alpha)R(\gamma)\varphi_i, \beta+\gamma)p)1 &= (-(\underline{m}, \alpha)\mathrm{R}(\gamma)\varphi_i, \beta+\gamma)p \\ ((0, \beta+\gamma)p)1 &= (0, \beta+\gamma)p \\ (((\underline{m}, \alpha)\varphi_i R(\beta), \beta+\gamma)p)1 &= ((\underline{m}, \alpha)\varphi_i R(\beta), \beta+\gamma)p \end{aligned}$$

and

$$\begin{aligned} ((-(\underline{m}, \alpha)R(\gamma)\varphi_i, \beta+\gamma)p)\pi &= (-(\underline{m}, \alpha)R(\gamma)\varphi_i, \beta+\gamma) \\ ((0, \beta+\gamma)p)\pi &= (0, \beta+\gamma) \\ (((\underline{m}, \alpha)\varphi_i R(\beta), \beta+\gamma)p)\pi &= ((\underline{m}, \alpha)\varphi_i R(\beta), \beta+\gamma). \end{aligned}$$

Now, using the definition of p in lemma 9, the action of $\{1, \pi\}$ on $T_{(\beta,\gamma)}$ may be rewritten

$$\begin{aligned} ((\underline{m}, \alpha)\varphi_i R(\gamma), \beta+\gamma)1 &= ((\underline{m}, \alpha)\varphi_i R(\gamma), \beta+\gamma) \\ (0, \beta+\gamma)1 &= (0, \beta+\gamma) \\ (-(\underline{m}, \alpha)R(\beta)\varphi_i, \beta+\gamma)1 &= (-(\underline{m}, \alpha)R(\beta)\varphi_i, \beta+\gamma) \end{aligned}$$

and

$$\begin{aligned} ((\underline{m}, \alpha)\varphi_i R(\gamma), \beta+\gamma)\pi &= (-(\underline{m}, \alpha)R(\gamma)\varphi_i, \beta+\gamma) \\ (0, \beta+\gamma)\pi &= (0, \beta+\gamma) \\ (-(\underline{m}, \alpha)R(\beta)\varphi_i, \beta+\gamma)\pi &= ((\underline{m}, \alpha)\varphi_i R(\beta), \beta+\gamma) \end{aligned}$$

for all $(\underline{m}, \alpha)$ in $\boldsymbol{Z}_N]S_3$, as in the statement of this lemma. To verify that the action of $\{1, \pi\}$ on $T_{(\beta,\gamma)}$ is faithful, note that

$$\begin{aligned} 1\colon\ T_{(\beta,\gamma)} &\rightarrow T_{(\beta,\gamma)};\ x \rightarrow x \\ \pi\colon\ T_{(\beta,\gamma)} &\rightarrow T_{(\beta,\gamma)};\ x \rightarrow x\pi \end{aligned}$$

are permutations of the set $T_{(\beta,\gamma)}$. If $\theta \in \{1, \pi\}$and θ belongs to the kernel of the action then $x\theta = x$ for all $x \in T_{(\beta,\gamma)}$. Then $\theta = 1$, since π moves $((\underline{0}, 1)\varphi_i R(\gamma), \beta+\gamma)$ to $(-(\underline{0}, 1)R(\gamma)\varphi_i, \beta+\gamma)$ which are distinct elements of $\boldsymbol{Z}(\boldsymbol{Z}_N]S_3)]\boldsymbol{Z}S_3$. □

CASE 2. Suppose $\beta = \gamma$. Note that in the labelling of the sectors of the sheets of the t-Riemann surface in figures 4 and 5, $R(\beta) = R(\gamma)$ and $\beta+\gamma = 2\beta$ in the group algebra $\boldsymbol{Z}S_3$.

13. LEMMA. *Referring to lemma 8, fix a common representative* φ_i *of the left and right cosets of* S_3 *in* $Sym(\mathbf{Z}_N]S_3)$. *Fix a pair* $(\beta, \beta)\in S_3\times S_3$. *Referring to lemma 5, define a subset* $T_{(\beta,\beta)}$ *of* $\mathbf{Z}(\mathbf{Z}_N]S_3)]\mathbf{Z}S_3$ *as follows:*

$$T_{(\beta,\beta)} = \left\{((\underline{m}, \alpha), 2\beta) | (\underline{m}, \alpha)\in \mathbf{Z}_N]S_3\right\} \cup \left\{(0, 2\beta)\right\} \cup \left\{(-(\underline{m}, \alpha), 2\beta) | (\underline{m}, \alpha)\in \mathbf{Z}_N]S_3\right\}.$$

Referring to the preceding discussion, consider the composite function

$$z \to t = z^2 \to w = t^{12} = z^{24}$$

of the complex z-plane to the w-Riemann surface. There is a copy of the set $T_{(\beta,\beta)}$ *on the upper sheet and a copy of the set* $T_{(\beta,\beta)}$ *on the lower sheet of the t-Riemann surface according to the labels of the sectors in figures* 4 *and* 5. *Note that in this case* $R(\beta) = R(\gamma)$ *and* $\beta+\gamma = 2\beta$ *with the branch point labelled by the element* $(0, 2\beta)$ *of both copies. The rotation of the z-plane by* π *radians induces a permutation*

$$p\colon T_{(\beta,\beta)} \to T_{(\beta,\beta)}$$

given by

$$\begin{aligned}(-(\underline{m}, \alpha)R(\beta)\varphi_i, 2\beta)p &= ((\underline{m}, \alpha)\varphi_i R(\beta), 2\beta)\\ (0, 2\beta)p &= (0, 2\beta)\\ ((\underline{m}, \alpha)\varphi_i R(\beta), 2\beta)p &= (-(\underline{m}, \alpha)R(\beta)\varphi_i, 2\beta)\end{aligned}$$

for all $(\underline{m}, \alpha)\in\mathbf{Z}_N]S_3$, *such that each point of the copy of* $T_{(\beta,\beta)}$ *on the upper sheet moves continuously along a circular curve that winds exactly once around the branch point, to the point superposed directly below it on the copy of* $T_{(\beta,\beta)}$ *on the lower sheet of the t-Riemann surface. Then* $p = p^{-1}$ *so that* $<p> = \{1, p\}$ *is a two-element cyclic subgroup of the full permutation group* $Sym(T_{(\beta,\beta)})$ *and* $<p>$ *acts faithfully on the set* $T_{(\beta,\beta)}$.

PROOF: As in the proof of lemma 9, $T_{(\beta,\beta)}$ is seen to be a well-defined subset of $\mathbf{Z}(\mathbf{Z}_N]S_3)]\mathbf{Z}S_3$ by setting the appropriate coefficients to zero in a typical element as described in lemma 5. Both $\varphi_i R(\beta)$, $R(\beta)\varphi_i$ are permutations of the set $\mathbf{Z}_N]S_3$ and the rotation of the z-plane by π radians clearly induces a permutation p of the set $T_{(\beta,\beta)}$ as described. Furthermore, it is clear from the definition that $p = p^{-1}$ by chasing elements of $T_{(\beta,\beta)}$. Then $<p> = \{1, p\}$ as a subgroup of $Sym(T_{(\beta,\beta)})$ and $<p>$ acts faithfully on the set $T_{(\beta,\beta)}$. □

14. LEMMA. *Referring to lemma* 13, *let* $1\colon \mathbf{C}\to\mathbf{C}$; $z\to z$ *denote the identity and* $\pi\colon \mathbf{C}\to\mathbf{C}$; $z\to -z$ *denote the rotation through an angle of* π *radians of the z-plane. Then there is a well-defined action of the two-element cyclic group* $\{1, \pi\}$ *on the set* $T_{(\beta,\beta)}$ *given by*

$$((\underline{m}, \alpha)\varphi_i R(\beta), 2\beta)1 = ((\underline{m}, \alpha)\varphi_i R(\beta), 2\beta)$$
$$(0, 2\beta)1 = (0, 2\beta)$$
$$(-(\underline{m}, \alpha)R(\beta)\varphi_i, 2\beta)1 = (-(\underline{m}, \alpha)R(\beta)\varphi_i, 2\beta)$$

and

$$((\underline{m}, \alpha)\varphi_i R(\beta), 2\beta)\pi = (-(\underline{m}, \alpha)R(\beta)\varphi_i, 2\beta)$$
$$(0, 2\beta)\pi = (0, 2\beta)$$
$$(-(\underline{m}, \alpha)R(\beta)\varphi_i, 2\beta)\pi = ((\underline{m}, \alpha)\varphi_i R(\beta), 2\beta)$$

for all $(\underline{m}, \alpha)$ *in* $\boldsymbol{Z}_N]S_3$. *This action is faithful.*

PROOF: The isomorphism $1 \to 1$, $p \to \pi$ of the two-element cyclic groups $\{1, p\}$ and $\{1, \pi\}$ establishes the lemma. □

VI. MAIN CONSTRUCTION

RÉSUMÉ. Let us review the final goal. Recall the definition made in section I. We have defined N to be the minimal number of colours required to properly colour any map from the class of all maps on the sphere. We know that $4 \leq N \leq 6$. We have chosen a specific map $\boldsymbol{m}(N)$ on the sphere which requires all of the N colours 0, 1, ..., N-1 to properly colour it. The map $\boldsymbol{m}(N)$ has been properly coloured and the regions of $\boldsymbol{m}(N)$ partitioned into disjoint, nonempty equivalence classes $\underline{0}$, $\underline{1}$, ..., $\underline{N\text{-}1}$ according to the colour they receive. The set $\{\underline{0}, \underline{1}, ..., \underline{N\text{-}1}\}$ is endowed with the structure of the cyclic group $\boldsymbol{Z}_N$ under addition modulo N. In section III we have built the split extension $\boldsymbol{Z}_N]S_3$. The underlying set $\boldsymbol{Z}_N]S_3$ of cardinality $6N$ is taken to be the point set of a Steiner system $S(N+1, 2N, 6N)$ which will be constructed in this section. We are required to define the blocks of size $2N$ and show that every set of N+1 points is contained in a unique block. Once this goal is achieved, lemma 3 shows that $N \leq 4$.

15. LEMMA. *Let* $\boldsymbol{Z}_N]S_3$ *denote the split extension defined in lemma* 4 *and* $Sym(\boldsymbol{Z}_N]S_3)$ *denote the full permutation group on the set* $\boldsymbol{Z}_N]S_3$. *Define*

$$\mu: Sym(\boldsymbol{Z}_N]S_3) \to Sym(\boldsymbol{Z}_N]S_3)$$

by

$$\psi = R(\gamma)\varphi_i \to \varphi_i R(\gamma) = \psi^{\mu}.$$

Then μ *is a bijection of the set* $Sym(\boldsymbol{Z}_N]S_3)$ *with itself.*

PROOF: Referring to lemma 8, μ is well-defined since each $\psi \in Sym(\boldsymbol{Z}_N]S_3)$ may be written uniquely as $\psi = R(\gamma)\varphi_i$ for some $\gamma \in S_3$ and some φ_i. Then μ is a surjection because for any $\psi \in Sym(\boldsymbol{Z}_N]S_3)$ one may also write $\psi = \varphi_i R(\gamma)$ uniquely for some $\gamma \in S_3$ and some φ_i, whence $R(\gamma)\varphi_i \to \varphi_i R(\gamma) = \psi$. Since $Sym(\boldsymbol{Z}_N]S_3)$ is a finite set, μ must be a bijection by counting. □

16. LEMMA. *Define the set G as follows:*

$$G = \left\{ \begin{pmatrix} \psi \\ \psi^{\mu} \end{pmatrix} \middle| \psi \in Sym(\mathbf{Z}_N]S_3) \right\} = \left\{ \begin{pmatrix} R(\gamma)\varphi_i \\ \varphi_i R(\gamma) \end{pmatrix} \middle| \begin{matrix} \gamma \in S_3 \\ i = 1, \ldots, k \end{matrix} \right\}.$$

Define multiplication in G as follows:

$$\begin{pmatrix} \psi_1 \\ \psi_1{}^{\mu} \end{pmatrix} \begin{pmatrix} \psi_2 \\ \psi_2{}^{\mu} \end{pmatrix} = \begin{pmatrix} (\psi_1\psi_2) \\ (\psi_1\psi_2)^{\mu} \end{pmatrix}$$

i.e.

$$\begin{pmatrix} R(\gamma_1)\varphi_{i_1} \\ \varphi_{i_1} R(\gamma_1) \end{pmatrix} \begin{pmatrix} R(\gamma_2)\varphi_{i_2} \\ \varphi_{i_2} R(\gamma_2) \end{pmatrix} = \begin{pmatrix} R(\gamma_3)\varphi_{i_3} \\ \varphi_{i_3} R(\gamma_3) \end{pmatrix}$$

where $R(\gamma_3)\varphi_{i_3}$ is the unique expression for $R(\gamma_1)\varphi_{i_1}R(\gamma_2)\varphi_{i_2}$ according to the right coset decomposition of S_3 in $Sym(\mathbf{Z}_N]S_3)$. Then G is a group.

PROOF: Referring to lemma 8 and lemma 15, the set G is well-defined by the decomposition of $Sym(\mathbf{Z}_N]S_3)$ into the left and right cosets of S_3 by the φ_i. Define

$$\mu' : Sym(\mathbf{Z}_N]S_3) \rightarrow G\ ;\ \psi \rightarrow \begin{pmatrix} \psi \\ \psi^{\mu} \end{pmatrix}.$$

Then μ' is a well-defined bijection of the set $Sym(\mathbf{Z}_N]S_3)$ with G since μ is a bijection by lemma 15. The definition of multiplication in G mirrors the multiplication in $Sym(\mathbf{Z}_N]S_3)$ via μ' and is designed to make G a group and μ' an isomorphism. □

17. LEMMA. *Consider the set $\mathbf{Z}_N]S_3$ and let*

$$\begin{pmatrix} \psi \\ \psi^{\mu} \end{pmatrix} = \begin{pmatrix} R(\gamma)\varphi_i \\ \varphi_i R(\gamma) \end{pmatrix} \in G.$$

Define

$$\uparrow \begin{pmatrix} \psi \\ \psi^{\mu} \end{pmatrix} : \mathbf{Z}_N]S_3 \rightarrow \mathbf{Z}_N]S_3 \text{ by}$$

$$(\underline{m}, \alpha) \rightarrow (\underline{m}, \alpha)\uparrow \begin{pmatrix} \psi \\ \psi^{\mu} \end{pmatrix} = (\underline{m}, \alpha)\uparrow \begin{pmatrix} R(\gamma)\varphi_i \\ \varphi_i R(\gamma) \end{pmatrix} = (\underline{m}, \alpha)R(\gamma)\varphi_i.$$

Define

$$\downarrow\begin{pmatrix}\psi\\ \psi^{\mu}\end{pmatrix}: \boldsymbol{Z}_N]S_3 \rightarrow \boldsymbol{Z}_N]S_3 \text{ by}$$

$$(\underline{m}, \alpha) \rightarrow (\underline{m}, \alpha)\downarrow\begin{pmatrix}\psi\\ \psi^{\mu}\end{pmatrix} = (\underline{m}, \alpha)\downarrow\begin{pmatrix}R(\gamma)\varphi_i\\ \varphi_i R(\gamma)\end{pmatrix} = (\underline{m}, \alpha)\varphi_i R(\gamma).$$

Then

$$(\underline{m}, \alpha)\uparrow\begin{pmatrix}\psi\\ \psi^{\mu}\end{pmatrix} = (\underline{m}, \alpha)\downarrow\begin{pmatrix}\psi\\ \psi^{\mu}\end{pmatrix} \text{ for all } (\underline{m}, \alpha)\in\boldsymbol{Z}_N]S_3.$$

Both ↑ and ↓ are well-defined, faithful and $|\boldsymbol{Z}_N]S_3|$*-transitive right actions of the group G on the set* $\boldsymbol{Z}_N]S_3$.

PROOF: Referring to lemma 12 and lemma 14, put $\beta = 1$. Working in the set $T_{(1,\gamma)}$, for each $(\underline{m}, \alpha)\in\boldsymbol{Z}_N]S_3$ we have

$$\left((\underline{m}, \alpha)\downarrow\begin{pmatrix}\psi\\ \psi^{\mu}\end{pmatrix}, 1+\gamma\right)$$
$$= \left((\underline{m}, \alpha)\downarrow\begin{pmatrix}R(\gamma)\varphi_i\\ \varphi_i R(\gamma)\end{pmatrix}, 1+\gamma\right)$$
$$= \left((\underline{m}, \alpha)\varphi_i R(\gamma), 1+\gamma\right)$$
$$= \left(-(\underline{m}, \alpha)R(\gamma)\varphi_i, 1+\gamma\right)\pi$$
$$= \left(-(\underline{m}, \alpha\gamma)\varphi_i, 1+\gamma\right)\pi$$
$$= \left(-(\underline{m}, \alpha\gamma)R(1)\varphi_i, 1+\gamma\right)\pi$$
$$= \left((\underline{m}, \alpha\gamma)\varphi_i R(1), 1+\gamma\right)$$
$$= \left((\underline{m}, \alpha\gamma)\varphi_i, 1+\gamma\right)$$
$$= \left((\underline{m}, \alpha)R(\gamma)\varphi_i, 1+\gamma\right)$$
$$= \left((\underline{m}, \alpha)\uparrow\begin{pmatrix}R(\gamma)\varphi_i\\ \varphi_i R(\gamma)\end{pmatrix}, 1+\gamma\right)$$
$$= \left((\underline{m}, \alpha)\uparrow\begin{pmatrix}\psi\\ \psi^{\mu}\end{pmatrix}, 1+\gamma\right)$$

using the action of the two-element cyclic group $\{1, \pi\}$ on the set $T_{(1,\gamma)}$ according to lemma 12 and lemma 14. Hence

$$(\underline{m}, \alpha)\uparrow\begin{pmatrix}\psi\\ \psi^{\mu}\end{pmatrix} = (\underline{m}, \alpha)\downarrow\begin{pmatrix}\psi\\ \psi^{\mu}\end{pmatrix} \quad \text{for all } (\underline{m}, \alpha)\in \mathbf{Z}_N]S_3.$$

Since the action $\uparrow$ is the usual action of $Sym(\mathbf{Z}_N]S_3)$ on the set $\mathbf{Z}_N]S_3$, it is faithful and $|\mathbf{Z}_N]S_3|$-transitive. By the last equality, so is the $\downarrow$ action. □

18. LEMMA. *Let* $(\underline{m}_1, \alpha_1), \ldots, (\underline{m}_r, \alpha_r)$ *be any r distinct elements of* $\mathbf{Z}_N]S_3$ *and let* $(\underline{n}_1, \beta_1), \ldots, (\underline{n}_s, \beta_s)$ *be any s distinct elements of* $\mathbf{Z}_N]S_3$. *Let*

$$H_{r,s} = \left\{ \begin{pmatrix}\psi\\ \psi^{\mu}\end{pmatrix}\in G \;\middle|\; \begin{array}{l} (\underline{m}_i, \alpha_i)\uparrow\begin{pmatrix}\psi\\ \psi^{\mu}\end{pmatrix} = (\underline{m}_i, \alpha_i) \text{ for } i = 1, \ldots, r \text{ and} \\ (\underline{n}_j, \beta_j)\downarrow\begin{pmatrix}\psi\\ \psi^{\mu}\end{pmatrix} = (\underline{n}_j, \beta_j) \text{ for } j = 1, \ldots, s \end{array} \right\}$$

then $H_{r,s}$ *is a subgroup of G.*

PROOF: Note that if $\psi = R(\gamma)\varphi_i = 1$ then $\varphi_i = R(\gamma)^{-1}$ so that $\psi^{\mu} = \varphi_i R(\gamma) = R(\gamma)^{-1}R(\gamma) = 1$. Then

$$\begin{pmatrix}1\\ 1^{\mu}\end{pmatrix}\in H_{r,s}$$

since

$$(\underline{m}_i, \alpha_i)\uparrow\begin{pmatrix}1\\ 1^{\mu}\end{pmatrix} = (\underline{m}_i, \alpha_i)1 = (\underline{m}_i, \alpha_i)$$

for $i = 1, \ldots, r$ and

$$(\underline{n}_j, \beta_j)\downarrow\begin{pmatrix}1\\ 1^{\mu}\end{pmatrix} = (\underline{n}_j, \beta_j)1^{\mu} = (\underline{n}_j, \beta_j)1 = (\underline{n}_j, \beta_j)$$

for $j = 1, \ldots, s$. If

$$\begin{pmatrix}\psi_1\\ \psi_1^{\mu}\end{pmatrix} \text{ and } \begin{pmatrix}\psi_2\\ \psi_2^{\mu}\end{pmatrix}\in H_{r,s}$$

then

$$(\underline{m}_i, \alpha_i)\uparrow\left(\begin{pmatrix}\psi_1\\ \psi_1^{\mu}\end{pmatrix}\begin{pmatrix}\psi_2\\ \psi_2^{\mu}\end{pmatrix}\right) = (\underline{m}_i, \alpha_i)\uparrow\begin{pmatrix}\psi_1\\ \psi_1^{\mu}\end{pmatrix}\uparrow\begin{pmatrix}\psi_2\\ \psi_2^{\mu}\end{pmatrix}$$

$$= (\underline{m}_i, \alpha_i)\uparrow\begin{pmatrix}\psi_2\\ \psi_2{}^{\mu}\end{pmatrix} = (\underline{m}_i, \alpha_i) \text{ for } i = 1, \ldots, r$$

and

$$(\underline{n}_j, \beta_j)\downarrow\left(\begin{pmatrix}\psi_1\\ \psi_1{}^{\mu}\end{pmatrix}\begin{pmatrix}\psi_2\\ \psi_2{}^{\mu}\end{pmatrix}\right) = (\underline{n}_j, \beta_j)\downarrow\begin{pmatrix}\psi_1\\ \psi_1{}^{\mu}\end{pmatrix}\downarrow\begin{pmatrix}\psi_2\\ \psi_2{}^{\mu}\end{pmatrix}$$

$$= (\underline{n}_j, \beta_j)\downarrow\begin{pmatrix}\psi_2\\ \psi_2{}^{\mu}\end{pmatrix} = (\underline{n}_j, \beta_j) \text{ for } j = 1, \ldots, s.$$

Hence

$$\begin{pmatrix}\psi_1\\ \psi_1{}^{\mu}\end{pmatrix}\begin{pmatrix}\psi_2\\ \psi_2{}^{\mu}\end{pmatrix}\in H_{r,\, s}$$

Since G is finite, $H_{r,\, s}$ is a subgroup of G. □

Note that $\boldsymbol{Z}_N$ is embedded as the subgroup $\{(\underline{m}, 1)|\underline{m}\in\boldsymbol{Z}_N\}$ in $\boldsymbol{Z}_N]S_3$ and S_3 is embedded as the subgroup $\{(\underline{0}, \alpha)|\alpha\in S_3\}$ in $\boldsymbol{Z}_N]S_3$. Since $\boldsymbol{Z}_N]S_3 = \boldsymbol{Z}_N\times S_3$ is the direct product of groups by lemma 4, both $\boldsymbol{Z}_N$ and S_3 are normal subgroups. Recall the notation

$$S_3 = <\sigma, \rho> = \{1, \rho, \rho^2, \sigma, \sigma\rho, \sigma\rho^2\}.$$

19. LEMMA. *Define*

$$H = \left\{ \begin{pmatrix}\psi\\ \psi^{\mu}\end{pmatrix}\in G \;\middle|\; \begin{matrix}(\underline{m}, 1)\uparrow\begin{pmatrix}\psi\\ \psi^{\mu}\end{pmatrix} = (\underline{m}, 1) \textit{ for all } \underline{m}\in\boldsymbol{Z}_N \textit{ and} \\ (\underline{0}, \sigma)\downarrow\begin{pmatrix}\psi\\ \psi^{\mu}\end{pmatrix} = (\underline{0}, \sigma)\end{matrix} \right\}.$$

Then given

$$\begin{pmatrix}\psi\\ \psi^{\mu}\end{pmatrix}\in H$$

either

$$(\underline{m}, \alpha)\downarrow\begin{pmatrix}\psi\\ \psi^{\mu}\end{pmatrix} = (\underline{m}, \alpha) \textit{ for all } (\underline{m}, \alpha)\in\boldsymbol{Z}_N]S_3$$

or

$$(\underline{m}, \alpha)\downarrow\begin{pmatrix}\psi \\ \psi^{\mu}\end{pmatrix} = (\underline{m}, \alpha^{\sigma}) \textit{ for all } (\underline{m}, \alpha) \in \mathbf{Z}_N]S_3.$$

PROOF: H is a well-defined subgroup of G according to lemma 18. Let

$$\begin{pmatrix}\psi \\ \psi^{\mu}\end{pmatrix} = \begin{pmatrix}R(\gamma)\varphi_i \\ \varphi_i R(\gamma)\end{pmatrix} \in H$$

and $(\underline{m}, \alpha) \in \mathbf{Z}_N]S_3$ be given. Referring to lemmas 12 and 14, put $\beta = \gamma^{-1}\alpha\gamma$. Working in the set $T_{(\beta,\gamma)}$ we have

$$\begin{aligned}
&\left((\underline{m}, \alpha)\downarrow\begin{pmatrix}\psi \\ \psi^{\mu}\end{pmatrix},\ \beta+\gamma\right) \\
&= \left((\underline{m}, \alpha)\downarrow\begin{pmatrix}R(\gamma)\varphi_i \\ \varphi_i R(\gamma)\end{pmatrix},\ \beta+\gamma\right) \\
&= \left((\underline{m}, \alpha)\varphi_i R(\gamma),\ \beta+\gamma\right) \\
&= \left(-(\underline{m}, \alpha)R(\gamma)\varphi_i,\ \beta+\gamma\right)\pi \\
&= \left(-(\underline{m}, \alpha\gamma)\varphi_i,\ \beta+\gamma\right)\pi \\
&= \left(-(\underline{m}, \gamma\beta)\varphi_i,\ \beta+\gamma\right)\pi \\
&= \left(-(\underline{m}, \gamma)R(\beta)\varphi_i,\ \beta+\gamma\right)\pi \\
&= \left((\underline{m}, \gamma)\varphi_i R(\beta),\ \beta+\gamma\right) \\
&= \left((\underline{m}, 1)R(\gamma)\varphi_i R(\beta),\ \beta+\gamma\right) \\
&= \left((\underline{m}, 1)R(\beta),\ \beta+\gamma\right) \\
&= \left((\underline{m}, \beta),\ \beta+\gamma\right)
\end{aligned}$$

using the definition of H and the action of the two-element cyclic group $\{1,\pi\}$ on the set $T_{(\beta,\gamma)}$. Hence

$$(\underline{m}, \alpha)\downarrow\begin{pmatrix}\psi \\ \psi^{\mu}\end{pmatrix} = (\underline{m}, \beta) = (\underline{m}, \gamma^{-1}\alpha\gamma) = (\underline{m}, \alpha^{\gamma}).$$

Now since

$$(\underline{0}, \sigma) = (\underline{0}, \sigma)\downarrow\begin{pmatrix}\psi \\ \psi^{\mu}\end{pmatrix} = (\underline{0}, \sigma^{\gamma})$$

by hypothesis, we have $\sigma = \sigma^{\gamma}$. Hence $\gamma\sigma = \sigma\gamma$ so that either $\gamma = 1$ or $\gamma = \sigma$. □

20. LEMMA. *Let H be the subgroup of G defined in lemma* 19. *Then H is a nontrivial* *group of involutions of the set* $\mathbf{Z}_N]S_3$. *In particular, every nontrivial element of H is of order* 2.

PROOF: Define

$$\psi\colon \mathbf{Z}_N]S_3 \to \mathbf{Z}_N]S_3;\ (\underline{m}, \alpha) \to (\underline{m}, \alpha^{\sigma}).$$

Then

$$(\underline{m}, 1)\uparrow\begin{pmatrix}\psi \\ \psi^{\mu}\end{pmatrix} = (\underline{m}, 1)\psi = (\underline{m}, 1^{\sigma}) = (\underline{m}, 1) \text{ for all } \underline{m}\in\mathbf{Z}_N$$

and

$$(\underline{0}, \sigma)\downarrow\begin{pmatrix}\psi \\ \psi^{\mu}\end{pmatrix} = (\underline{0}, \sigma)\uparrow\begin{pmatrix}\psi \\ \psi^{\mu}\end{pmatrix} = (\underline{0}, \sigma)\psi = (\underline{0}, \sigma^{\sigma}) = (\underline{0}, \sigma).$$

Now $\psi \neq 1$, so

$$\begin{pmatrix}1 \\ 1^{\mu}\end{pmatrix} \neq \begin{pmatrix}\psi \\ \psi^{\mu}\end{pmatrix} \in H,$$

hence H is nontrivial. To show that each nontrivial element of H is of order 2, let

$$\begin{pmatrix}\psi \\ \psi^{\mu}\end{pmatrix} = \begin{pmatrix}R(\gamma)\varphi_i \\ \varphi_i R(\gamma)\end{pmatrix} \in H.$$

Then by the proof of lemma 19, $\gamma = 1$ or $\gamma = \sigma$. In particular $\gamma^2 = 1$. Hence, for any $(\underline{m}, \alpha)\in\mathbf{Z}_N]S_3$

$$(\underline{m}, \alpha)\downarrow\begin{pmatrix}R(\gamma)\varphi_i \\ \varphi_i R(\gamma)\end{pmatrix}^2$$

$$= (\underline{m}, \alpha)\downarrow\begin{pmatrix}R(\gamma)\varphi_i \\ \varphi_i R(\gamma)\end{pmatrix}\downarrow\begin{pmatrix}R(\gamma)\varphi_i \\ \varphi_i R(\gamma)\end{pmatrix}$$

$$= (\underline{m}, \alpha^{\gamma})\downarrow\begin{pmatrix}R(\gamma)\varphi_i \\ \varphi_i R(\gamma)\end{pmatrix}$$

$$= (\underline{m}, (\alpha^{\gamma})^{\gamma}) = (\underline{m}, \alpha).$$

Since the $\downarrow$ action of G on the set $\mathbf{Z}_N]S_3$ is faithful,

$$\begin{pmatrix} R(\gamma)\varphi_i \\ \varphi_i R(\gamma) \end{pmatrix}^2 = \begin{pmatrix} 1 \\ 1^\mu \end{pmatrix}$$

the identity element of G. □

21. LEMMA. *Denote the right cosets of $\boldsymbol{Z}_N$ in $\boldsymbol{Z}_N]S_3$ by*

$$\boldsymbol{Z}_N,\ \boldsymbol{Z}_N\rho,\ \boldsymbol{Z}_N\rho^2,\ \boldsymbol{Z}_N\sigma,\ \boldsymbol{Z}_N\sigma\rho,\ \boldsymbol{Z}_N\sigma\rho^2.$$

Define

$$Fix{\downarrow}(H) = \left\{ (\underline{m}, \alpha)\in\boldsymbol{Z}_N]S_3 \,\middle|\, (\underline{m}, \alpha){\downarrow}\begin{pmatrix} \psi \\ \psi^\mu \end{pmatrix} = (\underline{m}, \alpha) \textit{ for all } \begin{pmatrix} \psi \\ \psi^\mu \end{pmatrix}\in H \right\}.$$

Then $Fix{\downarrow}(H) = \{(\underline{m}, \alpha)\in\boldsymbol{Z}_N]S_3 | \alpha = 1$ or $\alpha = \sigma\}$. The ↓ action of a nontrivial element of H transposes the coset $\boldsymbol{Z}_N\rho$ with the coset $\boldsymbol{Z}_N\rho^2$ and transposes the coset $\boldsymbol{Z}_N\sigma\rho$ with the coset $\boldsymbol{Z}_N\sigma\rho^2$.

PROOF: By lemmas 19 and 20, the elements

$$\begin{pmatrix} \psi \\ \psi^\mu \end{pmatrix}\in H$$

are of two kinds:

(i) $(\underline{m}, \alpha){\downarrow}\begin{pmatrix} \psi \\ \psi^\mu \end{pmatrix} = (\underline{m}, \alpha)$ for all $(\underline{m}, \alpha)\in\boldsymbol{Z}_N]S_3$

in which case

$$\begin{pmatrix} \psi \\ \psi^\mu \end{pmatrix} = \begin{pmatrix} 1 \\ 1^\mu \end{pmatrix}$$

the identity element of H, and

(ii) $(\underline{m}, \alpha){\downarrow}\begin{pmatrix} \psi \\ \psi^\mu \end{pmatrix} = (\underline{m}, \alpha^\sigma)$ for all $(\underline{m}, \alpha)\in\boldsymbol{Z}_N]S_3$

in which case

$$\begin{pmatrix} 1 \\ 1^\mu \end{pmatrix} \neq \begin{pmatrix} \psi \\ \psi^\mu \end{pmatrix}$$

is an element of order 2 in H. In the second case, compute according to the cosets of $\boldsymbol{Z}_N$ in $\boldsymbol{Z}_N]S_3$:

$$(\underline{m}, 1)\downarrow\begin{pmatrix}\psi\\ \psi^{\mu}\end{pmatrix} = (\underline{m}, 1^{\sigma}) = (\underline{m}, 1) \text{ for all } \underline{m}\in\boldsymbol{Z}_N$$

$$(\underline{m}, \sigma)\downarrow\begin{pmatrix}\psi\\ \psi^{\mu}\end{pmatrix} = (\underline{m}, \sigma^{\sigma}) = (\underline{m}, \sigma) \text{ for all } \underline{m}\in\boldsymbol{Z}_N$$

$$(\underline{m}, \rho)\downarrow\begin{pmatrix}\psi\\ \psi^{\mu}\end{pmatrix} = (\underline{m}, \rho^{\sigma}) = (\underline{m}, \rho^2) \text{ for all } \underline{m}\in\boldsymbol{Z}_N$$

$$(\underline{m}, \rho^2)\downarrow\begin{pmatrix}\psi\\ \psi^{\mu}\end{pmatrix} = (\underline{m}, (\rho^2)^{\sigma}) = (\underline{m}, \rho) \text{ for all } \underline{m}\in\boldsymbol{Z}_N$$

$$(\underline{m}, \sigma\rho)\downarrow\begin{pmatrix}\psi\\ \psi^{\mu}\end{pmatrix} = (\underline{m}, (\sigma\rho)^{\sigma}) = (\underline{m}, \sigma\rho^2) \text{ for all } \underline{m}\in\boldsymbol{Z}_N$$

$$(\underline{m}, \sigma\rho^2)\downarrow\begin{pmatrix}\psi\\ \psi^{\mu}\end{pmatrix} = (\underline{m}, (\sigma\rho^2)^{\sigma}) = (\underline{m}, \sigma\rho) \text{ for all } \underline{m}\in\boldsymbol{Z}_N$$

and the lemma follows. □

22. LEMMA. *Let* $Norm_G(H)$ *denote the normalizer of* H *in* G. *The action* ↓ *of* G *on* $\boldsymbol{Z}_N]S_3$ *restricts to an action* ↓ *of* $Norm_G(H)$ *on* $Fix\downarrow(H)$ *which is* $(|\boldsymbol{Z}_N|+1)$*-transitive.*

PROOF: Let

$$\begin{pmatrix}\psi\\ \psi^{\mu}\end{pmatrix}\in G.$$

First show that

$$Fix\downarrow\left(\begin{pmatrix}\psi\\ \psi^{\mu}\end{pmatrix}^{-1} H\begin{pmatrix}\psi\\ \psi^{\mu}\end{pmatrix}\right) = Fix\downarrow(H)\downarrow\begin{pmatrix}\psi\\ \psi^{\mu}\end{pmatrix}$$

as follows:

$$(\underline{m}, \alpha)\in Fix\downarrow\left(\begin{pmatrix}\psi\\ \psi^{\mu}\end{pmatrix}^{-1} H\begin{pmatrix}\psi\\ \psi^{\mu}\end{pmatrix}\right)$$

$$\Leftrightarrow (\underline{m}, \alpha)\downarrow\left(\begin{pmatrix}\psi\\ \psi^{\mu}\end{pmatrix}^{-1}\begin{pmatrix}\psi^*\\ \psi^{*\mu}\end{pmatrix}\begin{pmatrix}\psi\\ \psi^{\mu}\end{pmatrix}\right) = (\underline{m}, \alpha) \text{ for all } \begin{pmatrix}\psi^*\\ \psi^{*\mu}\end{pmatrix}\in H$$

$$\Leftrightarrow (\underline{m}, \alpha)\downarrow\left(\begin{pmatrix}\psi\\ \psi^{\mu}\end{pmatrix}^{-1}\begin{pmatrix}\psi^*\\ \psi^{*\mu}\end{pmatrix}\right) = (\underline{m}, \alpha)\downarrow\begin{pmatrix}\psi\\ \psi^{\mu}\end{pmatrix}^{-1} \text{ for all } \begin{pmatrix}\psi^*\\ \psi^{*\mu}\end{pmatrix}\in H$$

$$\Leftrightarrow (\underline{m}, \alpha)\downarrow\begin{pmatrix}\psi\\ \psi^{\mu}\end{pmatrix}^{-1}\downarrow\begin{pmatrix}\psi^{*}\\ \psi^{*\mu}\end{pmatrix} = (\underline{m}, \alpha)\downarrow\begin{pmatrix}\psi\\ \psi^{\mu}\end{pmatrix}^{-1} \text{ for all } \begin{pmatrix}\psi^{*}\\ \psi^{*\mu}\end{pmatrix}\in H$$

$$\Leftrightarrow (\underline{m}, \alpha)\downarrow\begin{pmatrix}\psi\\ \psi^{\mu}\end{pmatrix}^{-1}\in Fix\downarrow(H)$$

$$\Leftrightarrow (\underline{m}, \alpha)\in Fix\downarrow(H)\downarrow\begin{pmatrix}\psi\\ \psi^{\mu}\end{pmatrix}.$$

Now let

$$\begin{pmatrix}\psi\\ \psi^{\mu}\end{pmatrix}\in Norm_G(H) = \left\{\begin{pmatrix}\psi\\ \psi^{\mu}\end{pmatrix}\in G \,\middle|\, \begin{pmatrix}\psi\\ \psi^{\mu}\end{pmatrix}^{-1} H\begin{pmatrix}\psi\\ \psi^{\mu}\end{pmatrix} = H\right\}.$$

Then

$$Fix\downarrow(H) = Fix\downarrow\left(\begin{pmatrix}\psi\\ \psi^{\mu}\end{pmatrix}^{-1} H\begin{pmatrix}\psi\\ \psi^{\mu}\end{pmatrix}\right) = Fix\downarrow(H)\downarrow\begin{pmatrix}\psi\\ \psi^{\mu}\end{pmatrix}$$

showing that the action restricts to an action of $Norm_G(H)$ on $Fix\downarrow(H)$. Now to show that the action $\downarrow$ of $Norm_G(H)$ on $Fix\downarrow(H) = \boldsymbol{Z}_N \cup \boldsymbol{Z}_N\sigma$ is $(|\boldsymbol{Z}_N|+1)$-transitive, label the elements of $\boldsymbol{Z}_N$ as $(\underline{m}_1, \alpha_1)$, ..., $(\underline{m}_N, \alpha_N)$ and label $(\underline{0}, \sigma) = (\underline{m}_{N+1}, \alpha_{N+1})$. Let $(\underline{m}^*_1, \alpha^*_1)$, ..., $(\underline{m}^*_{N+1}, \alpha^*_{N+1})$ be any $|\boldsymbol{Z}_N|+1$ distinct points of $Fix\downarrow(H)$. It is enough to show that there exists

$$\begin{pmatrix}\psi\\ \psi^{\mu}\end{pmatrix}\in Norm_G(H)$$

such that

$$(\underline{m}^*_i, \alpha^*_i)\downarrow\begin{pmatrix}\psi\\ \psi^{\mu}\end{pmatrix} = (\underline{m}_i, \alpha_i) \text{ for } i = 1, \ldots, N+1.$$

Now there exists

$$\begin{pmatrix}\psi\\ \psi^{\mu}\end{pmatrix}\in G$$

such that

$$(\underline{m}^*_i, \alpha^*_i)\downarrow\begin{pmatrix}\psi\\ \psi^{\mu}\end{pmatrix} = (\underline{m}_i, \alpha_i) \text{ for } i = 1, \ldots, N+1.$$

Hence

$$(\underline{m}^*_i, \alpha^*_i) = (\underline{m}_i, \alpha_i)\downarrow\begin{pmatrix}\psi\\ \psi^{\mu}\end{pmatrix}^{-1} \text{ for } i = 1, ..., N+1.$$

Note that for every

$$\begin{pmatrix}\psi^*\\ \psi^{*\mu}\end{pmatrix}\in H$$

and for $i = 1, ..., N+1$:

$$(\underline{m}_i, \alpha_i)\downarrow\left(\begin{pmatrix}\psi\\ \psi^{\mu}\end{pmatrix}^{-1}\begin{pmatrix}\psi^*\\ \psi^{*\mu}\end{pmatrix}\begin{pmatrix}\psi\\ \psi^{\mu}\end{pmatrix}\right)$$

$$= (\underline{m}_i, \alpha_i)\downarrow\begin{pmatrix}\psi\\ \psi^{\mu}\end{pmatrix}^{-1}\downarrow\begin{pmatrix}\psi^*\\ \psi^{*\mu}\end{pmatrix}\downarrow\begin{pmatrix}\psi\\ \psi^{\mu}\end{pmatrix}$$

$$= (\underline{m}^*_i, \alpha^*_i)\downarrow\begin{pmatrix}\psi^*\\ \psi^{*\mu}\end{pmatrix}\downarrow\begin{pmatrix}\psi\\ \psi^{\mu}\end{pmatrix}$$

$$= (\underline{m}^*_i, \alpha^*_i)\downarrow\begin{pmatrix}\psi\\ \psi^{\mu}\end{pmatrix}$$

$$= (\underline{m}_i, \alpha_i).$$

Hence

$$\begin{pmatrix}\psi\\ \psi^{\mu}\end{pmatrix}^{-1}\begin{pmatrix}\psi^*\\ \psi^{*\mu}\end{pmatrix}\begin{pmatrix}\psi\\ \psi^{\mu}\end{pmatrix}\in H \text{ for all } \begin{pmatrix}\psi^*\\ \psi^{*\mu}\end{pmatrix}\in H \Rightarrow \begin{pmatrix}\psi\\ \psi^{\mu}\end{pmatrix}\in Norm_G(H). \ \Box$$

23. LEMMA. *There exists a Steiner system* $S(N+1, 2N, 6N)$, *where the points are the elements of the set* $\boldsymbol{Z}_N]S_3$ *and the set of blocks is*

$$\left\{ Fix\downarrow(H)\downarrow\begin{pmatrix}\psi\\ \psi^{\mu}\end{pmatrix} \middle| \begin{pmatrix}\psi\\ \psi^{\mu}\end{pmatrix}\in G \right\}.$$

PROOF: There are $6N = |\boldsymbol{Z}_N]S_3|$ points. Each block, for a fixed

$$\begin{pmatrix}\psi\\ \psi^{\mu}\end{pmatrix}\in G$$

contains

$$2N = |\boldsymbol{Z}_N \cup \boldsymbol{Z}_N\sigma| = |Fix{\downarrow}(H)| = \left| Fix{\downarrow}(H){\downarrow}\begin{pmatrix}\psi \\ \psi^{\mu}\end{pmatrix} \right|$$

points. Label the elements of $\boldsymbol{Z}_N$ as $(\underline{m}_1, \alpha_1), ..., (\underline{m}_N, \alpha_N)$ and label $(\underline{0}, \sigma) = (\underline{m}_{N+1}, \alpha_{N+1})$. Let $(\underline{m}^*_1, \alpha^*_1), ..., (\underline{m}^*_{N+1}, \alpha^*_{N+1})$ be any $|\boldsymbol{Z}_N|+1$ distinct points of $\boldsymbol{Z}_N]S_3$. Then there exists

$$\begin{pmatrix}\psi \\ \psi^{\mu}\end{pmatrix} \in G$$

such that

$$(\underline{m}_i, \alpha_i){\downarrow}\begin{pmatrix}\psi \\ \psi^{\mu}\end{pmatrix} = (\underline{m}^*_i, \alpha^*_i) \text{ for } i = 1, ..., N+1.$$

Hence, there is at least one block, namely $Fix{\downarrow}(H)$, that contains the points $(\underline{m}^*_1, \alpha^*_1), ..., (\underline{m}^*_{N+1}, \alpha^*_{N+1})$. It remains to show that this is the unique block that contains the points $(\underline{m}^*_1, \alpha^*_1), ..., (\underline{m}^*_{N+1}, \alpha^*_{N+1})$. Suppose $(\underline{m}^*_1, \alpha^*_1), ..., (\underline{m}^*_{N+1}, \alpha^*_{N+1})$ are contained in

$$Fix{\downarrow}(H){\downarrow}\begin{pmatrix}\psi^* \\ \psi^{*\mu}\end{pmatrix} \text{ for some } \begin{pmatrix}\psi^* \\ \psi^{*\mu}\end{pmatrix} \in G.$$

Then there exist points $(\underline{m}^{**}_1, \alpha^{**}_1), ..., (\underline{m}^{**}_{N+1}, \alpha^{**}_{N+1})$ in $Fix{\downarrow}(H)$ such that

$$(\underline{m}^*_i, \alpha^*_i) = (\underline{m}^{**}_i, \alpha^{**}_i){\downarrow}\begin{pmatrix}\psi^* \\ \psi^{*\mu}\end{pmatrix} \text{ for } i = 1, ..., N+1.$$

By lemma 22, there exists

$$\begin{pmatrix}\psi^{**} \\ \psi^{**\mu}\end{pmatrix} \in Norm_G(H)$$

such that

$$(\underline{m}^{**}_i, \alpha^{**}_i) = (\underline{m}_i, \alpha_i){\downarrow}\begin{pmatrix}\psi^{**} \\ \psi^{**\mu}\end{pmatrix} \text{ for } i = 1, ..., N+1.$$

Hence for $i = 1, ..., N+1$

$$(\underline{m}_i, \alpha_i){\downarrow}\begin{pmatrix}\psi \\ \psi^{\mu}\end{pmatrix}$$

$$= (\underline{m}^{*}_{i}, \alpha^{*}_{i})$$

$$= (\underline{m}^{**}_{i}, \alpha^{**}_{i})\downarrow\begin{pmatrix}\psi^{*}\\ \psi^{*\mu}\end{pmatrix}$$

$$= (\underline{m}_{i}, \alpha_{i})\downarrow\begin{pmatrix}\psi^{**}\\ \psi^{**\mu}\end{pmatrix}\downarrow\begin{pmatrix}\psi^{*}\\ \psi^{*\mu}\end{pmatrix}$$

$$\Rightarrow (\underline{m}_{i}, \alpha_{i}) = (\underline{m}_{i}, \alpha_{i})\downarrow\left(\begin{pmatrix}\psi^{**}\\ \psi^{**\mu}\end{pmatrix}\begin{pmatrix}\psi^{*}\\ \psi^{*\mu}\end{pmatrix}\begin{pmatrix}\psi\\ \psi^{\mu}\end{pmatrix}^{-1}\right) \text{ for } i = 1, \ldots, N+1.$$

Then by lemma 17

$$(\underline{m}_{i}, \alpha_{i}) = (\underline{m}_{i}, \alpha_{i})\uparrow\left(\begin{pmatrix}\psi^{**}\\ \psi^{**\mu}\end{pmatrix}\begin{pmatrix}\psi^{*}\\ \psi^{*\mu}\end{pmatrix}\begin{pmatrix}\psi\\ \psi^{\mu}\end{pmatrix}^{-1}\right) \text{ for } i = 1, \ldots, N$$

and

$$(\underline{m}_{N+1}, \alpha_{N+1}) = (\underline{m}_{N+1}, \alpha_{N+1})\downarrow\left(\begin{pmatrix}\psi^{**}\\ \psi^{**\mu}\end{pmatrix}\begin{pmatrix}\psi^{*}\\ \psi^{*\mu}\end{pmatrix}\begin{pmatrix}\psi\\ \psi^{\mu}\end{pmatrix}^{-1}\right).$$

Hence

$$\begin{pmatrix}\psi^{**}\\ \psi^{**\mu}\end{pmatrix}\begin{pmatrix}\psi^{*}\\ \psi^{*\mu}\end{pmatrix}\begin{pmatrix}\psi\\ \psi^{\mu}\end{pmatrix}^{-1} \in H.$$

Now H is a subgroup of $Norm_G(H)$

$$\Rightarrow \begin{pmatrix}\psi^{**}\\ \psi^{**\mu}\end{pmatrix}\begin{pmatrix}\psi^{*}\\ \psi^{*\mu}\end{pmatrix}\begin{pmatrix}\psi\\ \psi^{\mu}\end{pmatrix}^{-1} \in Norm_G(H)$$

$$\Rightarrow \begin{pmatrix}\psi^{*}\\ \psi^{*\mu}\end{pmatrix}\begin{pmatrix}\psi\\ \psi^{\mu}\end{pmatrix}^{-1} \in \begin{pmatrix}\psi^{**}\\ \psi^{**\mu}\end{pmatrix}^{-1} Norm_G(H) = Norm_G(H)$$

$$\Rightarrow \left(\begin{pmatrix}\psi^{*}\\ \psi^{*\mu}\end{pmatrix}\begin{pmatrix}\psi\\ \psi^{\mu}\end{pmatrix}^{-1}\right) H \left(\begin{pmatrix}\psi^{*}\\ \psi^{*\mu}\end{pmatrix}\begin{pmatrix}\psi\\ \psi^{\mu}\end{pmatrix}^{-1}\right)^{-1} = H$$

$$\Rightarrow \begin{pmatrix}\psi^{*}\\ \psi^{*\mu}\end{pmatrix}\begin{pmatrix}\psi\\ \psi^{\mu}\end{pmatrix}^{-1} H \begin{pmatrix}\psi\\ \psi^{\mu}\end{pmatrix}\begin{pmatrix}\psi^{*}\\ \psi^{*\mu}\end{pmatrix}^{-1} = H$$

$$\Rightarrow \begin{pmatrix}\psi\\ \psi^{\mu}\end{pmatrix}^{-1} H \begin{pmatrix}\psi\\ \psi^{\mu}\end{pmatrix} = \begin{pmatrix}\psi^{*}\\ \psi^{*\mu}\end{pmatrix}^{-1} H \begin{pmatrix}\psi^{*}\\ \psi^{*\mu}\end{pmatrix}.$$

Now, using the first fact in the proof of lemma 22

$$Fix{\downarrow}(H){\downarrow}\begin{pmatrix}\psi^*\\ \psi^{*\mu}\end{pmatrix}$$

$$= Fix{\downarrow}\left(\begin{pmatrix}\psi^*\\ \psi^{*\mu}\end{pmatrix}^{-1} H \begin{pmatrix}\psi^*\\ \psi^{*\mu}\end{pmatrix}\right)$$

$$= Fix{\downarrow}\left(\begin{pmatrix}\psi\\ \psi^{\mu}\end{pmatrix}^{-1} H \begin{pmatrix}\psi\\ \psi^{\mu}\end{pmatrix}\right)$$

$$= Fix{\downarrow}(H){\downarrow}\begin{pmatrix}\psi\\ \psi^{\mu}\end{pmatrix}.$$

This establishes the uniqueness of the block. □

24. THEOREM. *Any map on the sphere can be properly coloured by using at most four colours.*

PROOF: Referring to section I, we have defined N to be the minimal number of colours required to properly colour any map from the class of all maps on the sphere. Based on the definition of N, we have selected a specific map $\boldsymbol{m}(N)$ on the sphere which requires no fewer than N colours to be properly coloured. Based on the definition of the map $\boldsymbol{m}(N)$ we have selected a proper colouring of its regions using the N colours 0, 1, ..., N-1. Working with the fixed number N, the fixed map $\boldsymbol{m}(N)$, and the fixed proper colouring of the regions of the map $\boldsymbol{m}(N)$, lemma 23 has explicitly constructed a Steiner system $S(N+1, 2N, 6N)$. Now lemma 3 implies that N cannot exceed four. □

APPENDIX 1

COMMON SYSTEMS OF COSET REPRESENTATIVES

ASHAY DHARWADKER

ABSTRACT

Using the axiom of choice, we prove that given any group G and a finite subgroup H, there always exists a common system of coset representatives for the left and right cosets of H in G.

We shall prove that given any group G and a finite subgroup H, there always exists a common system of coset representatives for the left and right cosets of H in G. Precise definitions and examples are given below. The proof uses the standard von Neumann - Bernays - Gödel (NBG) axioms of set theory [1] together with

The Axiom of Choice. Given any set X of nonempty pairwise disjoint sets, there is a set Y, called a *choice set*, that contains exactly one element of each set in X.

A nonempty set I together with a binary relation $\leq$ is called a *partially ordered set* if, for all i, j, k in I

- $i \leq i$ (*reflexivity*)
- $i \leq j$ and $j \leq k$ implies $i \leq k$ (*transitivity*)
- $i \leq j$ and $j \leq i$ implies $i = j$ (*antisymmetry*)

We write $i < j$ when $i \leq j$ and i is not equal to j. Given a nonempty subset J of a partially ordered set I, an element j_0 of J is called a *least element of* J if $j_0 \leq j$ for all j in J. A partially ordered set I is said to be *well-ordered* if every nonempty subset of I has a least element. Note that in a well-ordered set any two elements i, j are comparable since the subset $\{ i, j \}$ must have a least element. We shall use

The Well-Ordering Principle. Every set can be well-ordered.

Proof. See [1], the proof of proposition 4.37. The axiom of choice implies Zorn's lemma. Zorn's lemma implies the well-ordering principle. □

In particular, given any set X, we may index the elements of X by a well-ordered index set I and write $X = \{ x_i \mid i \text{ in } I \}$. In this notation we may now state and prove

The Transfinite Induction Principle. Let $X = \{ x_i \mid i \text{ in } I \}$ be any set indexed by a well-ordered set I. If P is a property such that, for any i in I, whenever all x_j with $j < i$ have property P, then x_i has property P, then all elements of X have property P.

Proof. Let $Y = \{ x \text{ in } X \mid x \text{ has property } P \}$. Suppose $X - Y$ is nonempty, then there is a least element x_i in $X - Y$. By the definition of least element and $X - Y$ we must have, for any x_j with $j < i$, that x_j has the property P. But then, by hypothesis, x_i has property P, a contradiction. Therefore, $X - Y$ is empty and $X = Y$. □

A set G together with a binary operation (written here in the usual multiplicative notation) is called a *group* if

- For all x, y, z in G, $x(yz) = (xy)z$ (*associativity*)
- There exists an *identity* element 1 in G such that for all x in G, $x1 = x = 1x$
- For each x in G, there exists an *inverse* element x^{-1} in G such that $xx^{-1} = 1 = x^{-1}x$

It is easy to show that the identity element 1 is unique and, for each x in G, the inverse element x^{-1} is unique, see [2]. A nonempty subset H of a group G is called a *subgroup* if, for all h_1, h_2 in H

- h_1h_2 is in H
- h_1^{-1} is in H

From the definition it follows that the identity element $1 = h_1h_1^{-1}$ is in H and the subgroup H is itself a group under the induced binary operation of multiplication. For any element x of G, the map $g \to xgx^{-1}$ is a bijection from G to G, called the *inner automorphism of G under conjugation by x* and this map induces a bijection from H to xHx^{-1} which is also a subgroup of G. Given any element x of G, the set $xH = \{ xh \mid h \text{ in } H \}$ is called a *left coset* of H in G, the set $Hx = \{ hx \mid h \text{ in } H \}$ is called a *right coset* of H in G and the set $HxH = \{ h_1xh_2 \mid h_1, h_2 \text{ in } H \}$ is called a *double coset* of H in G. An element of a coset is called a *representative* for that coset. The maps $h \to xh$ and $h \to hx$ induce bijections from H to xH and Hx respectively. Suppose z belongs to the left cosets xH and yH, then $z = xh_1 = yh_2$ for some h_1, h_2 in H, so $xH = yh_2h_1^{-1}H = yH$. Also, any x in G belongs to a left coset, namely xH. Thus G is the disjoint union of the left cosets of H. Similarly, G is the disjoint union of the right cosets of H and lemma 1 below proves that G is the disjoint union of the double cosets of H. Since $(Hx)^{-1} = \{ (hx)^{-1} \mid h \text{ in } H, x \text{ in } G \} = \{ x^{-1}h^{-1} \mid h \text{ in } H, x \text{ in } G \} = x^{-1}H$ and $(yH)^{-1} = \{ (yh)^{-1} \mid h \text{ in } H, y \text{ in } G \} = \{ h^{-1}y^{-1} \mid h \text{ in } H, y \text{ in } G \} = Hy^{-1}$, there is a bijection between the set of all left cosets and the set of all right cosets of H in G. A set consisting of exactly one representative of each left coset from the set of all left cosets of H in G is called a *system of representatives for the left cosets* of H in G. Similarly, a set consisting of exactly one representative of each right coset from the set of all right cosets of H in G is called a *system of representatives for the right cosets* of H in G. By the axiom of choice, a system of representatives for the left cosets of H in G exists and a system of representatives for the right cosets of H in G exists. A set that is simultaneously a system of representatives for the left cosets of H in G and a system of representatives for the right cosets of H in G is called a *common system of representatives for the left and right cosets* of H in G.

Lemma 1. Let G be a group and H a subgroup. Then G is the disjoint union of the set of double cosets $\{ HgH \mid g \text{ in } G \}$.

Proof. Suppose x belongs to the double cosets Hg_1H and Hg_2H. Then $x = h'_1g_1h'_2 = h''_1g_2h''_2$

for some h'_1, h'_2, h''_1, h''_2 in H. Then $g_1 = h'_1{}^{-1}h''_1g_2h''_2h'_2{}^{-1}$ and so, for any h_1, h_2 in H, we have $h_1g_1h_2 = h_1h'_1{}^{-1}h''_1g_2h''_2h'_2{}^{-1}h_2$ showing that Hg_1H is contained in Hg_2H. Similarly $g_2 = h''_1{}^{-1}h'_1g_1h'_2h''_2{}^{-1}$ and so, for any h_1, h_2 in H, we have $h_1g_2h_2 = h_1h''_1{}^{-1}h'_1g_1h'_2h''_2{}^{-1}h_2$ showing that Hg_2H is contained in Hg_1H. Thus $Hg_1H = Hg_2H$. This proves that distinct double cosets cannot have any elements in common and must be disjoint. Since every g in G can be written as $g = 1g1$, every g in G belongs to at least one double coset, namely HgH. This proves that the union of the disjoint double cosets is all of G. □

Lemma 2. Let G be a group and H a subgroup. Let HgH be a fixed double coset of H in G. Then

- Every left coset of H in G is either contained in HgH or disjoint from it. Hence HgH is the disjoint union of the left cosets of H in G that are contained in HgH.
- Every right coset of H in G is either contained in HgH or disjoint from it. Hence HgH is the disjoint union of the right cosets of H in G that are contained in HgH.

Proof. Let xH be a left coset of H in G. Suppose xh is an element of xH such that xh belongs to HgH. Then $xh = h_1gh_2$ for some h_1, h_2 in H, so $x = h_1gh_2h^{-1}$. Thus, for any h' in H, $xh' = h_1gh_2h^{-1}h'$ showing that the left coset xH is contained in HgH. This proves that either the left coset xH is contained in HgH or disjoint from it. Any two left cosets are disjoint because if x is in yH and zH then $x = yh_1= zh_2$ for some h_1, h_2 in H, so $z = yh_1h_2{}^{-1}$ shows that $yH = zH$. Also, every h_1gh_2 in HgH belongs to some left coset contained in HgH, namely h_1gH. This proves that HgH is the disjoint union of the left cosets of H in G that are contained in HgH. Similarly, let Hx be a right coset of H in G. Suppose hx is an element of Hx such that hx belongs to HgH. Then $hx = h_1gh_2$ for some h_1, h_2 in H, so $x = h^{-1}h_1gh_2$. Thus, for any h' in H, $h'x = h'h^{-1}h_1gh_2$ showing that the right coset Hx is contained in HgH. This proves that either the right coset Hx is contained in HgH or disjoint from it. Any two right cosets are disjoint because if x is in Hy and Hz then $x = h_1y = h_2z$ for some h_1, h_2 in H, so $z = h_2{}^{-1}h_1y$ shows that $Hy = Hz$. Also, every h_1gh_2 in HgH belongs to some right coset contained in HgH, namely Hgh_2. This proves that HgH is the disjoint union of the right cosets of H in G that are contained in HgH. □

Lemma 3. Let G be a group and H a finite subgroup. Let HgH be a fixed double coset of H in G. Then there exists a system of representatives for the left cosets of H in G that are contained in HgH such that distinct representatives belong to distinct right cosets of H in G that are contained in HgH.

Proof. There are two cases.

- **Case 1.** Suppose $Hg = gH$. Then HgH contains exactly one left coset $gH = HgH$ and exactly one right coset $Hg = HgH$. In this case, select g as a representative of the left coset gH and then g belongs to the unique right coset Hg contained in HgH.
- **Case 2.** Suppose Hg is not equal to gH. By lemma 2 and the well-ordering principle, let $\{ L_i \mid i \text{ in } I \}$ denote the set of left cosets of H in G that are contained in HgH, indexed by a well-ordered set I. Note that any left coset xH contained in HgH can be written as $xH = hgH$ for some h in H. Hence, by the axiom of choice, we can select $\{ h_i \text{ in } H \mid i \text{ in } I \}$ such that $\{ L_i \mid i \text{ in } I \} = \{ h_igH \mid i \text{ in } I \}$. We shall now use the principle of transfinite induction. Given i in I, assume that for all $j < i$ we have selected h'_j in H such that the right cosets Hgh'_j are all distinct. We claim that we can

select h'_i in H such that the right coset Hgh'_i is distinct from all the right cosets Hgh'_j where $j < i$. Suppose not. Then for each h in H there exists a right coset $Hgh'_j = Hgh$ with $j < i$. Thus for each h in H, there exist h'_j, h''_j, h'''_j in H such that $h''_j gh'_j = h'''_j gh$. That is, for each h in H, there exist h'_j, h''_j, h'''_j in H such that $h'''^{-1}_j h''_j g = ghh'^{-1}_j$. Thus Hg contains $gHh'^{-1}_j = gH$ and so H contains gHg^{-1}. *This is the point in the proof where we use the fact that H is finite.* Since the inner automorphism under conjugation by g is bijective and H is finite, $H = gHg^{-1}$. But then, $Hg = gH$, a contradiction to the assumption of case 2. Hence, our claim is true: we can select h'_i in H such that the right coset Hgh'_i is distinct from all the right cosets Hgh'_j where $j < i$. By the principle of transfinite induction, we can select distinct right cosets $\{ Hgh'_i \mid i \text{ in } I \}$. Note that $h_i gH = h_i gh'_i H$ and $Hgh'_i = Hh_i gh'_i$ for all i in I. Thus, the element $h_i gh'_i$ is a common representative for the left coset $h_i gH$ and the right coset Hgh'_i for all i in I. It follows that the set $\{ h_i gh'_i \mid i \text{ in } I \}$ is a system of representatives for the left cosets of H in G that are contained in HgH such that distinct representatives belong to distinct right cosets of H in G that are contained in HgH. □

Proposition. Let G be a group and H a finite subgroup. Then there exists a common system of coset representatives for the left and right cosets of H in G.

Proof. By lemma 1 and the axiom of choice, select a set $\{ g_j \text{ in } G \mid j \text{ in } J \}$ such that $\{ Hg_jH \mid j \text{ in } J \}$ is the set of disjoint double cosets whose union is G. By lemma 3 and the axiom of choice, select a set $\{ S_j \mid j \text{ in } J \}$ where S_j is a set of representatives for the left cosets of H in G that are contained in Hg_jH such that distinct representatives belong to distinct right cosets of H in G that are contained in Hg_jH. Form the union S of all the sets in $\{ S_j \mid j \text{ in } J \}$. Then S is a system of representatives for all left cosets of H in G such that distinct representatives belong to distinct right cosets of H in G. But, as observed above, there is a bijection between the set of all left cosets of H in G and the set of all right cosets of H in G. Thus each right coset of H in G must have an element in S. It follows that the set S must be a common system of representatives for the left and right cosets of H in G. □

Example 1. Let $G = S_3$ denote the *symmetric group on three letters* consisting of all permutations of the set $\{1, 2, 3\}$

$$1 = \begin{pmatrix} 1 & 2 & 3 \\ 1 & 2 & 3 \end{pmatrix} \quad \alpha = \begin{pmatrix} 1 & 2 & 3 \\ 2 & 3 & 1 \end{pmatrix} \quad \beta = \begin{pmatrix} 1 & 2 & 3 \\ 3 & 1 & 2 \end{pmatrix}$$

$$\gamma = \begin{pmatrix} 1 & 2 & 3 \\ 1 & 3 & 2 \end{pmatrix} \quad \delta = \begin{pmatrix} 1 & 2 & 3 \\ 3 & 2 & 1 \end{pmatrix} \quad \varepsilon = \begin{pmatrix} 1 & 2 & 3 \\ 2 & 1 & 3 \end{pmatrix}$$

together with the binary operation of permutation multiplication. To facilitate our computation, let us write the multiplication table for the group G explicitly:

	1	α	β	γ	δ	ε
1	1	α	β	γ	δ	ε
α	α	β	1	δ	ε	γ
β	β	1	α	ε	γ	δ
γ	γ	ε	δ	1	β	α
δ	δ	γ	ε	α	1	β
ε	ε	δ	γ	β	α	1

Consider the subgroup $H = \{ 1, \varepsilon \}$. The double cosets of H in G are $\{ 1, \varepsilon \}$ and $\{ \alpha, \beta, \gamma, \delta \}$. The left cosets of H in G are $\{ 1, \varepsilon \}$, $\{ \alpha, \gamma \}$ and $\{ \beta, \delta \}$. The right cosets of H in G are $\{ 1, \varepsilon \}$, $\{ \alpha, \delta \}$ and $\{ \beta, \gamma \}$. We may select ε as a common representative of the left coset $\{ 1, \varepsilon \}$ and the right coset $\{ 1, \varepsilon \}$ contained in the double coset $\{ 1, \varepsilon \}$. We may select γ and δ as common representatives for the left cosets $\{ \alpha, \gamma \}$, $\{ \beta, \delta \}$ and right cosets $\{ \alpha, \delta \}$, $\{ \beta, \gamma \}$ respectively, contained in the double coset $\{ \alpha, \beta, \gamma, \delta \}$. The union of the selected representatives $\{ \varepsilon, \gamma, \delta \}$ is a common system of representatives for the left and right cosets of H in G in this example where H is finite.

Example 2. Finally, we give an example of a group G and subgroup H that do not satisfy the hypotheses of the proposition and for which there cannot exist a common system of representatives for the left and right cosets of H in G. Consider the group G generated by x, y subject to the relation $xy = y^2x$. Let H be the subgroup generated by y. Then $H = \{ y^n \mid n \text{ is any integer} \}$ is an *infinite subgroup*. Using the relation inductively, it is easy to see that for any integer n, $xy^nx^{-1} = y^{2n}$. Thus the subgroup $xHx^{-1} = \{ y^{2n} \mid n \text{ is any integer} \}$ is properly contained in the subgroup H. This implies that the left coset xH is properly contained in the right coset Hx which is equal to the double coset HxH. But then by lemma 2, the double coset HxH contains at least two left cosets and exactly one right coset. Thus, it is impossible to select representatives for the left cosets of H in HxH that belong to distinct right cosets of H in HxH. By lemma 1 and lemma 2, it follows that it is impossible to select representatives for the left cosets of H in G that belong to distinct right cosets of H in G. Thus, there cannot exist a common system of representatives for the left and right cosets of H in G in this example where H is infinite.

APPENDIX 2

RIEMANN SURFACES

ASHAY DHARWADKER

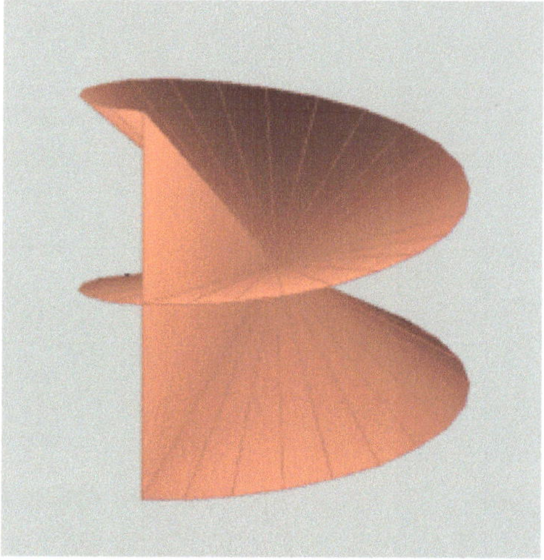

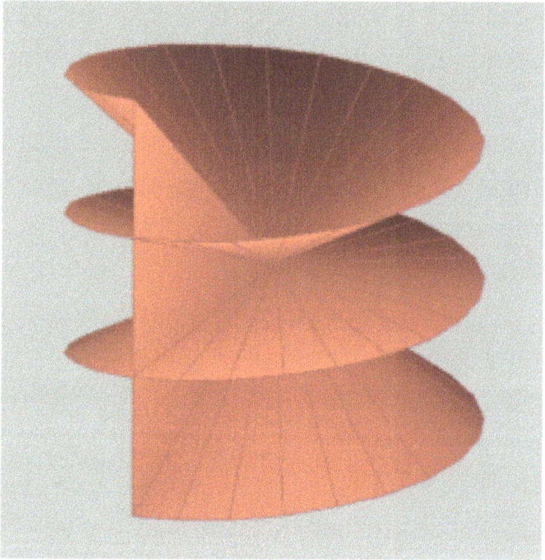

Riemann surfaces were first studied by Bernhard Riemann in his Inauguraldissertation at Göttingen in 1851.

Consider the function from the complex plane to itself given by $w = f(z) = z^n$, where n is at least 2. The z-plane may be divided into n sectors given by $arg\ z$ lying between $(k - 1)(2\pi/n)$ and $k(2\pi/n)$ for $k = 1, ..., n$. There is a one-to-one correspondence between each sector and the whole w-plane, except for the positive real axis. The image of each sector is obtained by performing a cut along the positive real axis; this cut has an upper and a lower edge. Corresponding to the n sectors in the z-plane, take n identical copies of the w-plane with the cut. These will be the *sheets* of the Riemann surface and are distinguished by a label k which serves to identify the corresponding sector. For $k = 1, ..., n$-1 attach the lower edge of the sheet labeled k with the upper edge of the sheet labeled $k + 1$. To complete the cycle, attach the lower edge of the sheet labeled n to the upper edge of the sheet labeled 1. In a physical sense, this is not possible without self-intersection but the idealized model shall be free of this discrepancy. The result of the construction is a *Riemann surface* whose points are in one-to-one correspondence with the points of the z-plane. This correspondence is continuous in the following sense. When z moves in its plane the corresponding point w is free to move on the Riemann surface. The point $w = 0$ connects all the sheets and is called the *branch point*. A curve must wind n times around the branch point before it closes. Now consider the n-valued relation $z = n^{th}$ *root* (w). To each nonzero w, there correspond n values of z. If the w-plane is replaced by the Riemann surface just constructed, then each complex nonzero w is represented by n points of the Riemann surface at superposed positions. Let the point on the uppermost sheet represent the principal value and the other n - 1 points represent the other values. Then $z = n^{th}$ *root* (w) becomes a single-valued, continuous, one-to-one correspondence of the points of the Riemann surface with the points of the z-plane. The Riemann surface is orientable, since every orientation of a sheet is carried over to the sheet next to it.

APPENDIX 3

THE WITT DESIGN

ASHAY DHARWADKER

A posteriori the proof of the four colour theorem (lemmas 1-23), we know that the Steiner system $S(N+1, 2N, 6N)$ has been constructed and $N = 4$. Then we can return to the beginning and use the lemmas 1-23 to construct the Steiner system $S(5, 8, 24)$ explicitly.

- There are 24 points.
- Each block consists of 8 points.
- Any 5 points are contained in a unique block.

By lemma 4, the 24 points of the Steiner system $S(5, 8, 24)$ are the elements of the underlying set $\boldsymbol{Z}_4]S_3$:

$(\underline{0}, 1)$	$(\underline{0}, \rho)$	$(\underline{0}, \rho^2)$	$(\underline{0}, \sigma\rho^2)$	$(\underline{0}, \sigma\rho)$	$(\underline{0}, \sigma)$
$(\underline{1}, 1)$	$(\underline{1}, \rho)$	$(\underline{1}, \rho^2)$	$(\underline{1}, \sigma\rho^2)$	$(\underline{1}, \sigma\rho)$	$(\underline{1}, \sigma)$
$(\underline{2}, 1)$	$(\underline{2}, \rho)$	$(\underline{2}, \rho^2)$	$(\underline{2}, \sigma\rho^2)$	$(\underline{2}, \sigma\rho)$	$(\underline{2}, \sigma)$
$(\underline{3}, 1)$	$(\underline{3}, \rho)$	$(\underline{3}, \rho^2)$	$(\underline{3}, \sigma\rho^2)$	$(\underline{3}, \sigma\rho)$	$(\underline{3}, \sigma)$

Using the definitions in the proof and lemma 23, the blocks of the Steiner system $S(5, 8, 24)$ are defined by the set

$$\left\{ Fix\downarrow(H)\downarrow\begin{pmatrix}\psi\\ \psi^{\mu}\end{pmatrix} \middle| \begin{pmatrix}\psi\\ \psi^{\mu}\end{pmatrix}\in G \right\}$$

There are 759 blocks as follows:

$$\mathbf{B_1} = \left\{ (\underline{0}, 1), (\underline{1}, 1), (\underline{2}, 1), (\underline{3}, 1), (\underline{0}, \sigma), (\underline{1}, \sigma), (\underline{2}, \sigma), (\underline{3}, \sigma) \right\}$$

$$\mathbf{B_2} = \left\{ (\underline{0}, 1), (\underline{1}, 1), (\underline{2}, 1), (\underline{3}, 1), (\underline{0}, \rho), (\underline{2}, \rho), (\underline{0}, \sigma\rho), (\underline{3}, \sigma\rho) \right\}$$

$$\mathbf{B_3} = \left\{ (\underline{0}, 1), (\underline{1}, 1), (\underline{2}, 1), (\underline{3}, 1), (\underline{1}, \rho), (\underline{3}, \rho), (\underline{1}, \sigma\rho), (\underline{2}, \sigma\rho) \right\}$$

$$\mathbf{B_4} = \left\{ (\underline{0}, 1), (\underline{1}, 1), (\underline{2}, 1), (\underline{3}, 1), (\underline{0}, \rho^2), (\underline{0}, \sigma\rho^2), (\underline{1}, \sigma\rho^2), (\underline{3}, \sigma\rho^2) \right\}$$

$$\mathbf{B_5} = \left\{ (\underline{0}, 1), (\underline{1}, 1), (\underline{2}, 1), (\underline{3}, 1), (\underline{1}, \rho^2), (\underline{2}, \rho^2), (\underline{3}, \rho^2), (\underline{2}, \sigma\rho^2) \right\}$$

$$\mathbf{B}_6 = \left\{ (\underline{0}, 1), (\underline{1}, 1), (\underline{2}, 1), (\underline{0}, \sigma), (\underline{0}, \rho), (\underline{2}, \sigma\rho), (\underline{2}, \rho^2), (\underline{1}, \sigma\rho^2) \right\}$$

$$\mathbf{B}_7 = \left\{ (\underline{0}, 1), (\underline{1}, 1), (\underline{2}, 1), (\underline{0}, \sigma), (\underline{1}, \rho), (\underline{2}, \rho), (\underline{0}, \rho^2), (\underline{3}, \rho^2) \right\}$$

$$\mathbf{B}_8 = \left\{ (\underline{0}, 1), (\underline{1}, 1), (\underline{2}, 1), (\underline{0}, \sigma), (\underline{3}, \rho), (\underline{3}, \sigma\rho), (\underline{1}, \rho^2), (\underline{3}, \sigma\rho^2) \right\}$$

$$\mathbf{B}_9 = \left\{ (\underline{0}, 1), (\underline{1}, 1), (\underline{2}, 1), (\underline{0}, \sigma), (\underline{0}, \sigma\rho), (\underline{1}, \sigma\rho), (\underline{0}, \sigma\rho^2), (\underline{2}, \sigma\rho^2) \right\}$$

$$\mathbf{B}_{10} = \left\{ (\underline{0}, 1), (\underline{1}, 1), (\underline{2}, 1), (\underline{1}, \sigma), (\underline{0}, \rho), (\underline{1}, \sigma\rho), (\underline{0}, \rho^2), (\underline{1}, \rho^2) \right\}$$

$$\mathbf{B}_{11} = \left\{ (\underline{0}, 1), (\underline{1}, 1), (\underline{2}, 1), (\underline{1}, \sigma), (\underline{1}, \rho), (\underline{3}, \sigma\rho), (\underline{2}, \rho^2), (\underline{0}, \sigma\rho^2) \right\}$$

$$\mathbf{B}_{12} = \left\{ (\underline{0}, 1), (\underline{1}, 1), (\underline{2}, 1), (\underline{1}, \sigma), (\underline{2}, \rho), (\underline{3}, \rho), (\underline{1}, \sigma\rho^2), (\underline{2}, \sigma\rho^2) \right\}$$

$$\mathbf{B}_{13} = \left\{ (\underline{0}, 1), (\underline{1}, 1), (\underline{2}, 1), (\underline{1}, \sigma), (\underline{0}, \sigma\rho), (\underline{2}, \sigma\rho), (\underline{3}, \rho^2), (\underline{3}, \sigma\rho^2) \right\}$$

$$\mathbf{B}_{14} = \left\{ (\underline{0}, 1), (\underline{1}, 1), (\underline{2}, 1), (\underline{2}, \sigma), (\underline{0}, \rho), (\underline{1}, \rho), (\underline{2}, \sigma\rho^2), (\underline{3}, \sigma\rho^2) \right\}$$

$$\mathbf{B}_{15} = \left\{ (\underline{0}, 1), (\underline{1}, 1), (\underline{2}, 1), (\underline{2}, \sigma), (\underline{2}, \rho), (\underline{2}, \sigma\rho), (\underline{1}, \rho^2), (\underline{0}, \sigma\rho^2) \right\}$$

$$\mathbf{B}_{16} = \left\{ (\underline{0}, 1), (\underline{1}, 1), (\underline{2}, 1), (\underline{2}, \sigma), (\underline{3}, \rho), (\underline{0}, \sigma\rho), (\underline{0}, \rho^2), (\underline{2}, \rho^2) \right\}$$

$$\mathbf{B}_{17} = \left\{ (\underline{0}, 1), (\underline{1}, 1), (\underline{2}, 1), (\underline{2}, \sigma), (\underline{1}, \sigma\rho), (\underline{3}, \sigma\rho), (\underline{3}, \rho^2), (\underline{1}, \sigma\rho^2) \right\}$$

$$\mathbf{B}_{18} = \left\{ (\underline{0}, 1), (\underline{1}, 1), (\underline{2}, 1), (\underline{3}, \sigma), (\underline{0}, \rho), (\underline{3}, \rho), (\underline{3}, \rho^2), (\underline{0}, \sigma\rho^2) \right\}$$

$$\mathbf{B}_{19} = \left\{ (\underline{0}, 1), (\underline{1}, 1), (\underline{2}, 1), (\underline{3}, \sigma), (\underline{1}, \rho), (\underline{0}, \sigma\rho), (\underline{1}, \rho^2), (\underline{1}, \sigma\rho^2) \right\}$$

$$\mathbf{B}_{20} = \left\{ (\underline{0}, 1), (\underline{1}, 1), (\underline{2}, 1), (\underline{3}, \sigma), (\underline{2}, \rho), (\underline{1}, \sigma\rho), (\underline{2}, \rho^2), (\underline{3}, \sigma\rho^2) \right\}$$

$$\mathbf{B}_{21} = \left\{ (\underline{0}, 1), (\underline{1}, 1), (\underline{2}, 1), (\underline{3}, \sigma), (\underline{2}, \sigma\rho), (\underline{3}, \sigma\rho), (\underline{0}, \rho^2), (\underline{2}, \sigma\rho^2) \right\}$$

$$\mathbf{B}_{22} = \left\{ (\underline{0}, 1), (\underline{1}, 1), (\underline{3}, 1), (\underline{0}, \sigma), (\underline{0}, \rho), (\underline{1}, \sigma\rho), (\underline{3}, \rho^2), (\underline{3}, \sigma\rho^2) \right\}$$

$$\mathbf{B}_{23} = \left\{ (\underline{0}, 1), (\underline{1}, 1), (\underline{3}, 1), (\underline{0}, \sigma), (\underline{1}, \rho), (\underline{3}, \sigma\rho), (\underline{1}, \sigma\rho^2), (\underline{2}, \sigma\rho^2) \right\}$$

$$\mathbf{B}_{24} = \left\{ (\underline{0}, 1), (\underline{1}, 1), (\underline{3}, 1), (\underline{0}, \sigma), (\underline{2}, \rho), (\underline{3}, \rho), (\underline{2}, \rho^2), (\underline{0}, \sigma\rho^2) \right\}$$

$$\mathbf{B}_{25} = \left\{ (\underline{0}, 1), (\underline{1}, 1), (\underline{3}, 1), (\underline{0}, \sigma), (\underline{0}, \sigma\rho), (\underline{2}, \sigma\rho), (\underline{0}, \rho^2), (\underline{1}, \rho^2) \right\}$$

$$\mathbf{B}_{26} = \left\{ (\underline{0}, 1), (\underline{1}, 1), (\underline{3}, 1), (\underline{1}, \sigma), (\underline{0}, \rho), (\underline{2}, \sigma\rho), (\underline{0}, \sigma\rho^2), (\underline{2}, \sigma\rho^2) \right\}$$

$$\mathbf{B}_{27} = \left\{ (\underline{0}, 1), (\underline{1}, 1), (\underline{3}, 1), (\underline{1}, \sigma), (\underline{1}, \rho), (\underline{2}, \rho), (\underline{1}, \rho^2), (\underline{3}, \sigma\rho^2) \right\}$$

$$\mathbf{B_{28}} = \left\{ (\underline{0}, 1), (\underline{1}, 1), (\underline{3}, 1), (\underline{1}, \sigma), (\underline{3}, \rho), (\underline{3}, \sigma\rho), (\underline{0}, \rho^2), (\underline{3}, \rho^2) \right\}$$

$$\mathbf{B_{29}} = \left\{ (\underline{0}, 1), (\underline{1}, 1), (\underline{3}, 1), (\underline{1}, \sigma), (\underline{0}, \sigma\rho), (\underline{1}, \sigma\rho), (\underline{2}, \rho^2), (\underline{1}, \sigma\rho^2) \right\}$$

$$\mathbf{B_{30}} = \left\{ (\underline{0}, 1), (\underline{1}, 1), (\underline{3}, 1), (\underline{2}, \sigma), (\underline{0}, \rho), (\underline{3}, \rho), (\underline{1}, \rho^2), (\underline{1}, \sigma\rho^2) \right\}$$

$$\mathbf{B_{31}} = \left\{ (\underline{0}, 1), (\underline{1}, 1), (\underline{3}, 1), (\underline{2}, \sigma), (\underline{1}, \rho), (\underline{0}, \sigma\rho), (\underline{3}, \rho^2), (\underline{0}, \sigma\rho^2) \right\}$$

$$\mathbf{B_{32}} = \left\{ (\underline{0}, 1), (\underline{1}, 1), (\underline{3}, 1), (\underline{2}, \sigma), (\underline{2}, \rho), (\underline{1}, \sigma\rho), (\underline{0}, \rho^2), (\underline{2}, \sigma\rho^2) \right\}$$

$$\mathbf{B_{33}} = \left\{ (\underline{0}, 1), (\underline{1}, 1), (\underline{3}, 1), (\underline{2}, \sigma), (\underline{2}, \sigma\rho), (\underline{3}, \sigma\rho), (\underline{2}, \rho^2), (\underline{3}, \sigma\rho^2) \right\}$$

$$\mathbf{B_{34}} = \left\{ (\underline{0}, 1), (\underline{1}, 1), (\underline{3}, 1), (\underline{3}, \sigma), (\underline{0}, \rho), (\underline{1}, \rho), (\underline{0}, \rho^2), (\underline{2}, \rho^2) \right\}$$

$$\mathbf{B_{35}} = \left\{ (\underline{0}, 1), (\underline{1}, 1), (\underline{3}, 1), (\underline{3}, \sigma), (\underline{2}, \rho), (\underline{2}, \sigma\rho), (\underline{3}, \rho^2), (\underline{1}, \sigma\rho^2) \right\}$$

$$\mathbf{B_{36}} = \left\{ (\underline{0}, 1), (\underline{1}, 1), (\underline{3}, 1), (\underline{3}, \sigma), (\underline{3}, \rho), (\underline{0}, \sigma\rho), (\underline{2}, \sigma\rho^2), (\underline{3}, \sigma\rho^2) \right\}$$

$$\mathbf{B_{37}} = \left\{ (\underline{0}, 1), (\underline{1}, 1), (\underline{3}, 1), (\underline{3}, \sigma), (\underline{1}, \sigma\rho), (\underline{3}, \sigma\rho), (\underline{1}, \rho^2), (\underline{0}, \sigma\rho^2) \right\}$$

$$\mathbf{B_{38}} = \left\{ (\underline{0}, 1), (\underline{1}, 1), (\underline{0}, \sigma), (\underline{1}, \sigma), (\underline{0}, \rho), (\underline{1}, \rho), (\underline{3}, \rho), (\underline{0}, \sigma\rho) \right\}$$

$$\mathbf{B_{39}} = \left\{ (\underline{0}, 1), (\underline{1}, 1), (\underline{0}, \sigma), (\underline{1}, \sigma), (\underline{2}, \rho), (\underline{1}, \sigma\rho), (\underline{2}, \sigma\rho), (\underline{3}, \sigma\rho) \right\}$$

$$\mathbf{B_{40}} = \left\{ (\underline{0}, 1), (\underline{1}, 1), (\underline{0}, \sigma), (\underline{1}, \sigma), (\underline{0}, \rho^2), (\underline{2}, \rho^2), (\underline{2}, \sigma\rho^2), (\underline{3}, \sigma\rho^2) \right\}$$

$$\mathbf{B_{41}} = \left\{ (\underline{0}, 1), (\underline{1}, 1), (\underline{0}, \sigma), (\underline{1}, \sigma), (\underline{1}, \rho^2), (\underline{3}, \rho^2), (\underline{0}, \sigma\rho^2), (\underline{1}, \sigma\rho^2) \right\}$$

$$\mathbf{B_{42}} = \left\{ (\underline{0}, 1), (\underline{1}, 1), (\underline{0}, \sigma), (\underline{2}, \sigma), (\underline{0}, \rho), (\underline{3}, \sigma\rho), (\underline{0}, \rho^2), (\underline{0}, \sigma\rho^2) \right\}$$

$$\mathbf{B_{43}} = \left\{ (\underline{0}, 1), (\underline{1}, 1), (\underline{0}, \sigma), (\underline{2}, \sigma), (\underline{1}, \rho), (\underline{1}, \sigma\rho), (\underline{1}, \rho^2), (\underline{2}, \rho^2) \right\}$$

$$\mathbf{B_{44}} = \left\{ (\underline{0}, 1), (\underline{1}, 1), (\underline{0}, \sigma), (\underline{2}, \sigma), (\underline{2}, \rho), (\underline{0}, \sigma\rho), (\underline{1}, \sigma\rho^2), (\underline{3}, \sigma\rho^2) \right\}$$

$$\mathbf{B_{45}} = \left\{ (\underline{0}, 1), (\underline{1}, 1), (\underline{0}, \sigma), (\underline{2}, \sigma), (\underline{3}, \rho), (\underline{2}, \sigma\rho), (\underline{3}, \rho^2), (\underline{2}, \sigma\rho^2) \right\}$$

$$\mathbf{B_{46}} = \left\{ (\underline{0}, 1), (\underline{1}, 1), (\underline{0}, \sigma), (\underline{3}, \sigma), (\underline{0}, \rho), (\underline{2}, \rho), (\underline{1}, \rho^2), (\underline{2}, \sigma\rho^2) \right\}$$

$$\mathbf{B_{47}} = \left\{ (\underline{0}, 1), (\underline{1}, 1), (\underline{0}, \sigma), (\underline{3}, \sigma), (\underline{1}, \rho), (\underline{2}, \sigma\rho), (\underline{0}, \sigma\rho^2), (\underline{3}, \sigma\rho^2) \right\}$$

$$\mathbf{B_{48}} = \left\{ (\underline{0}, 1), (\underline{1}, 1), (\underline{0}, \sigma), (\underline{3}, \sigma), (\underline{3}, \rho), (\underline{1}, \sigma\rho), (\underline{0}, \rho^2), (\underline{1}, \sigma\rho^2) \right\}$$

$$\mathbf{B_{49}} = \left\{ (\underline{0}, 1), (\underline{1}, 1), (\underline{0}, \sigma), (\underline{3}, \sigma), (\underline{0}, \sigma\rho), (\underline{3}, \sigma\rho), (\underline{2}, \rho^2), (\underline{3}, \rho^2) \right\}$$

$$\mathbf{B}_{50} = \left\{ (\underline{0}, 1), (\underline{1}, 1), (\underline{1}, \sigma), (\underline{2}, \sigma), (\underline{0}, \rho), (\underline{2}, \rho), (\underline{2}, \rho^2), (\underline{3}, \rho^2) \right\}$$

$$\mathbf{B}_{51} = \left\{ (\underline{0}, 1), (\underline{1}, 1), (\underline{1}, \sigma), (\underline{2}, \sigma), (\underline{1}, \rho), (\underline{2}, \sigma\rho), (\underline{0}, \rho^2), (\underline{1}, \sigma\rho^2) \right\}$$

$$\mathbf{B}_{52} = \left\{ (\underline{0}, 1), (\underline{1}, 1), (\underline{1}, \sigma), (\underline{2}, \sigma), (\underline{3}, \rho), (\underline{1}, \sigma\rho), (\underline{0}, \sigma\rho^2), (\underline{3}, \sigma\rho^2) \right\}$$

$$\mathbf{B}_{53} = \left\{ (\underline{0}, 1), (\underline{1}, 1), (\underline{1}, \sigma), (\underline{2}, \sigma), (\underline{0}, \sigma\rho), (\underline{3}, \sigma\rho), (\underline{1}, \rho^2), (\underline{2}, \sigma\rho^2) \right\}$$

$$\mathbf{B}_{54} = \left\{ (\underline{0}, 1), (\underline{1}, 1), (\underline{1}, \sigma), (\underline{3}, \sigma), (\underline{0}, \rho), (\underline{3}, \sigma\rho), (\underline{1}, \sigma\rho^2), (\underline{3}, \sigma\rho^2) \right\}$$

$$\mathbf{B}_{55} = \left\{ (\underline{0}, 1), (\underline{1}, 1), (\underline{1}, \sigma), (\underline{3}, \sigma), (\underline{1}, \rho), (\underline{1}, \sigma\rho), (\underline{3}, \rho^2), (\underline{2}, \sigma\rho^2) \right\}$$

$$\mathbf{B}_{56} = \left\{ (\underline{0}, 1), (\underline{1}, 1), (\underline{1}, \sigma), (\underline{3}, \sigma), (\underline{2}, \rho), (\underline{0}, \sigma\rho), (\underline{0}, \rho^2), (\underline{0}, \sigma\rho^2) \right\}$$

$$\mathbf{B}_{57} = \left\{ (\underline{0}, 1), (\underline{1}, 1), (\underline{1}, \sigma), (\underline{3}, \sigma), (\underline{3}, \rho), (\underline{2}, \sigma\rho), (\underline{1}, \rho^2), (\underline{2}, \rho^2) \right\}$$

$$\mathbf{B}_{58} = \left\{ (\underline{0}, 1), (\underline{1}, 1), (\underline{2}, \sigma), (\underline{3}, \sigma), (\underline{0}, \rho), (\underline{0}, \sigma\rho), (\underline{1}, \sigma\rho), (\underline{2}, \sigma\rho) \right\}$$

$$\mathbf{B}_{59} = \left\{ (\underline{0}, 1), (\underline{1}, 1), (\underline{2}, \sigma), (\underline{3}, \sigma), (\underline{1}, \rho), (\underline{2}, \rho), (\underline{3}, \rho), (\underline{3}, \sigma\rho) \right\}$$

$$\mathbf{B}_{60} = \left\{ (\underline{0}, 1), (\underline{1}, 1), (\underline{2}, \sigma), (\underline{3}, \sigma), (\underline{0}, \rho^2), (\underline{1}, \rho^2), (\underline{3}, \rho^2), (\underline{3}, \sigma\rho^2) \right\}$$

$$\mathbf{B}_{61} = \left\{ (\underline{0}, 1), (\underline{1}, 1), (\underline{2}, \sigma), (\underline{3}, \sigma), (\underline{2}, \rho^2), (\underline{0}, \sigma\rho^2), (\underline{1}, \sigma\rho^2), (\underline{2}, \sigma\rho^2) \right\}$$

$$\mathbf{B}_{62} = \left\{ (\underline{0}, 1), (\underline{1}, 1), (\underline{0}, \rho), (\underline{1}, \rho), (\underline{2}, \rho), (\underline{1}, \sigma\rho), (\underline{0}, \sigma\rho^2), (\underline{1}, \sigma\rho^2) \right\}$$

$$\mathbf{B}_{63} = \left\{ (\underline{0}, 1), (\underline{1}, 1), (\underline{0}, \rho), (\underline{1}, \rho), (\underline{2}, \sigma\rho), (\underline{3}, \sigma\rho), (\underline{1}, \rho^2), (\underline{3}, \rho^2) \right\}$$

$$\mathbf{B}_{64} = \left\{ (\underline{0}, 1), (\underline{1}, 1), (\underline{0}, \rho), (\underline{2}, \rho), (\underline{3}, \rho), (\underline{2}, \sigma\rho), (\underline{0}, \rho^2), (\underline{3}, \sigma\rho^2) \right\}$$

$$\mathbf{B}_{65} = \left\{ (\underline{0}, 1), (\underline{1}, 1), (\underline{0}, \rho), (\underline{3}, \rho), (\underline{1}, \sigma\rho), (\underline{3}, \sigma\rho), (\underline{2}, \rho^2), (\underline{2}, \sigma\rho^2) \right\}$$

$$\mathbf{B}_{66} = \left\{ (\underline{0}, 1), (\underline{1}, 1), (\underline{0}, \rho), (\underline{0}, \sigma\rho), (\underline{0}, \rho^2), (\underline{3}, \rho^2), (\underline{1}, \sigma\rho^2), (\underline{2}, \sigma\rho^2) \right\}$$

$$\mathbf{B}_{67} = \left\{ (\underline{0}, 1), (\underline{1}, 1), (\underline{0}, \rho), (\underline{0}, \sigma\rho), (\underline{1}, \rho^2), (\underline{2}, \rho^2), (\underline{0}, \sigma\rho^2), (\underline{3}, \sigma\rho^2) \right\}$$

$$\mathbf{B}_{68} = \left\{ (\underline{0}, 1), (\underline{1}, 1), (\underline{1}, \rho), (\underline{2}, \rho), (\underline{0}, \sigma\rho), (\underline{2}, \sigma\rho), (\underline{2}, \rho^2), (\underline{2}, \sigma\rho^2) \right\}$$

$$\mathbf{B}_{69} = \left\{ (\underline{0}, 1), (\underline{1}, 1), (\underline{1}, \rho), (\underline{3}, \rho), (\underline{0}, \rho^2), (\underline{1}, \rho^2), (\underline{0}, \sigma\rho^2), (\underline{2}, \sigma\rho^2) \right\}$$

$$\mathbf{B}_{70} = \left\{ (\underline{0}, 1), (\underline{1}, 1), (\underline{1}, \rho), (\underline{3}, \rho), (\underline{2}, \rho^2), (\underline{3}, \rho^2), (\underline{1}, \sigma\rho^2), (\underline{3}, \sigma\rho^2) \right\}$$

$$\mathbf{B}_{71} = \left\{ (\underline{0}, 1), (\underline{1}, 1), (\underline{1}, \rho), (\underline{0}, \sigma\rho), (\underline{1}, \sigma\rho), (\underline{3}, \sigma\rho), (\underline{0}, \rho^2), (\underline{3}, \sigma\rho^2) \right\}$$

$\mathbf{B}_{72} = \{ (\underline{0}, 1), (\underline{1}, 1), (\underline{2}, \rho), (\underline{3}, \rho), (\underline{0}, \sigma\rho), (\underline{1}, \sigma\rho), (\underline{1}, \rho^2), (\underline{3}, \rho^2) \}$

$\mathbf{B}_{73} = \{ (\underline{0}, 1), (\underline{1}, 1), (\underline{2}, \rho), (\underline{3}, \sigma\rho), (\underline{0}, \rho^2), (\underline{1}, \rho^2), (\underline{2}, \rho^2), (\underline{1}, \sigma\rho^2) \}$

$\mathbf{B}_{74} = \{ (\underline{0}, 1), (\underline{1}, 1), (\underline{2}, \rho), (\underline{3}, \sigma\rho), (\underline{3}, \rho^2), (\underline{0}, \sigma\rho^2), (\underline{2}, \sigma\rho^2), (\underline{3}, \sigma\rho^2) \}$

$\mathbf{B}_{75} = \{ (\underline{0}, 1), (\underline{1}, 1), (\underline{3}, \rho), (\underline{0}, \sigma\rho), (\underline{2}, \sigma\rho), (\underline{3}, \sigma\rho), (\underline{0}, \sigma\rho^2), (\underline{1}, \sigma\rho^2) \}$

$\mathbf{B}_{76} = \{ (\underline{0}, 1), (\underline{1}, 1), (\underline{1}, \sigma\rho), (\underline{2}, \sigma\rho), (\underline{0}, \rho^2), (\underline{2}, \rho^2), (\underline{3}, \rho^2), (\underline{0}, \sigma\rho^2) \}$

$\mathbf{B}_{77} = \{ (\underline{0}, 1), (\underline{1}, 1), (\underline{1}, \sigma\rho), (\underline{2}, \sigma\rho), (\underline{1}, \rho^2), (\underline{1}, \sigma\rho^2), (\underline{2}, \sigma\rho^2), (\underline{3}, \sigma\rho^2) \}$

$\mathbf{B}_{78} = \{ (\underline{0}, 1), (\underline{2}, 1), (\underline{3}, 1), (\underline{0}, \sigma), (\underline{0}, \rho), (\underline{3}, \rho), (\underline{0}, \rho^2), (\underline{2}, \sigma\rho^2) \}$

$\mathbf{B}_{79} = \{ (\underline{0}, 1), (\underline{2}, 1), (\underline{3}, 1), (\underline{0}, \sigma), (\underline{1}, \rho), (\underline{0}, \sigma\rho), (\underline{2}, \rho^2), (\underline{3}, \sigma\rho^2) \}$

$\mathbf{B}_{80} = \{ (\underline{0}, 1), (\underline{2}, 1), (\underline{3}, 1), (\underline{0}, \sigma), (\underline{2}, \rho), (\underline{1}, \sigma\rho), (\underline{1}, \rho^2), (\underline{1}, \sigma\rho^2) \}$

$\mathbf{B}_{81} = \{ (\underline{0}, 1), (\underline{2}, 1), (\underline{3}, 1), (\underline{0}, \sigma), (\underline{2}, \sigma\rho), (\underline{3}, \sigma\rho), (\underline{3}, \rho^2), (\underline{0}, \sigma\rho^2) \}$

$\mathbf{B}_{82} = \{ (\underline{0}, 1), (\underline{2}, 1), (\underline{3}, 1), (\underline{1}, \sigma), (\underline{0}, \rho), (\underline{1}, \rho), (\underline{3}, \rho^2), (\underline{1}, \sigma\rho^2) \}$

$\mathbf{B}_{83} = \{ (\underline{0}, 1), (\underline{2}, 1), (\underline{3}, 1), (\underline{1}, \sigma), (\underline{2}, \rho), (\underline{2}, \sigma\rho), (\underline{0}, \rho^2), (\underline{2}, \rho^2) \}$

$\mathbf{B}_{84} = \{ (\underline{0}, 1), (\underline{2}, 1), (\underline{3}, 1), (\underline{1}, \sigma), (\underline{3}, \rho), (\underline{0}, \sigma\rho), (\underline{1}, \rho^2), (\underline{0}, \sigma\rho^2) \}$

$\mathbf{B}_{85} = \{ (\underline{0}, 1), (\underline{2}, 1), (\underline{3}, 1), (\underline{1}, \sigma), (\underline{1}, \sigma\rho), (\underline{3}, \sigma\rho), (\underline{2}, \sigma\rho^2), (\underline{3}, \sigma\rho^2) \}$

$\mathbf{B}_{86} = \{ (\underline{0}, 1), (\underline{2}, 1), (\underline{3}, 1), (\underline{2}, \sigma), (\underline{0}, \rho), (\underline{1}, \sigma\rho), (\underline{2}, \rho^2), (\underline{0}, \sigma\rho^2) \}$

$\mathbf{B}_{87} = \{ (\underline{0}, 1), (\underline{2}, 1), (\underline{3}, 1), (\underline{2}, \sigma), (\underline{1}, \rho), (\underline{3}, \sigma\rho), (\underline{0}, \rho^2), (\underline{1}, \rho^2) \}$

$\mathbf{B}_{88} = \{ (\underline{0}, 1), (\underline{2}, 1), (\underline{3}, 1), (\underline{2}, \sigma), (\underline{2}, \rho), (\underline{3}, \rho), (\underline{3}, \rho^2), (\underline{3}, \sigma\rho^2) \}$

$\mathbf{B}_{89} = \{ (\underline{0}, 1), (\underline{2}, 1), (\underline{3}, 1), (\underline{2}, \sigma), (\underline{0}, \sigma\rho), (\underline{2}, \sigma\rho), (\underline{1}, \sigma\rho^2), (\underline{2}, \sigma\rho^2) \}$

$\mathbf{B}_{90} = \{ (\underline{0}, 1), (\underline{2}, 1), (\underline{3}, 1), (\underline{3}, \sigma), (\underline{0}, \rho), (\underline{2}, \sigma\rho), (\underline{1}, \rho^2), (\underline{3}, \sigma\rho^2) \}$

$\mathbf{B}_{91} = \{ (\underline{0}, 1), (\underline{2}, 1), (\underline{3}, 1), (\underline{3}, \sigma), (\underline{1}, \rho), (\underline{2}, \rho), (\underline{0}, \sigma\rho^2), (\underline{2}, \sigma\rho^2) \}$

$\mathbf{B}_{92} = \{ (\underline{0}, 1), (\underline{2}, 1), (\underline{3}, 1), (\underline{3}, \sigma), (\underline{3}, \rho), (\underline{3}, \sigma\rho), (\underline{2}, \rho^2), (\underline{1}, \sigma\rho^2) \}$

$\mathbf{B}_{93} = \{ (\underline{0}, 1), (\underline{2}, 1), (\underline{3}, 1), (\underline{3}, \sigma), (\underline{0}, \sigma\rho), (\underline{1}, \sigma\rho), (\underline{0}, \rho^2), (\underline{3}, \rho^2) \}$

$$\mathbf{B}_{94} = \left\{ (\underline{0}, 1), (\underline{2}, 1), (\underline{0}, \sigma), (\underline{1}, \sigma), (\underline{0}, \rho), (\underline{2}, \rho), (\underline{0}, \sigma\rho^2), (\underline{3}, \sigma\rho^2) \right\}$$

$$\mathbf{B}_{95} = \left\{ (\underline{0}, 1), (\underline{2}, 1), (\underline{0}, \sigma), (\underline{1}, \sigma), (\underline{1}, \rho), (\underline{2}, \sigma\rho), (\underline{1}, \rho^2), (\underline{2}, \sigma\rho^2) \right\}$$

$$\mathbf{B}_{96} = \left\{ (\underline{0}, 1), (\underline{2}, 1), (\underline{0}, \sigma), (\underline{1}, \sigma), (\underline{3}, \rho), (\underline{1}, \sigma\rho), (\underline{2}, \rho^2), (\underline{3}, \rho^2) \right\}$$

$$\mathbf{B}_{97} = \left\{ (\underline{0}, 1), (\underline{2}, 1), (\underline{0}, \sigma), (\underline{1}, \sigma), (\underline{0}, \sigma\rho), (\underline{3}, \sigma\rho), (\underline{0}, \rho^2), (\underline{1}, \sigma\rho^2) \right\}$$

$$\mathbf{B}_{98} = \left\{ (\underline{0}, 1), (\underline{2}, 1), (\underline{0}, \sigma), (\underline{2}, \sigma), (\underline{0}, \rho), (\underline{0}, \sigma\rho), (\underline{1}, \rho^2), (\underline{3}, \rho^2) \right\}$$

$$\mathbf{B}_{99} = \left\{ (\underline{0}, 1), (\underline{2}, 1), (\underline{0}, \sigma), (\underline{2}, \sigma), (\underline{1}, \rho), (\underline{3}, \rho), (\underline{0}, \sigma\rho^2), (\underline{1}, \sigma\rho^2) \right\}$$

$$\mathbf{B}_{100} = \left\{ (\underline{0}, 1), (\underline{2}, 1), (\underline{0}, \sigma), (\underline{2}, \sigma), (\underline{2}, \rho), (\underline{3}, \sigma\rho), (\underline{2}, \rho^2), (\underline{2}, \sigma\rho^2) \right\}$$

$$\mathbf{B}_{101} = \left\{ (\underline{0}, 1), (\underline{2}, 1), (\underline{0}, \sigma), (\underline{2}, \sigma), (\underline{1}, \sigma\rho), (\underline{2}, \sigma\rho), (\underline{0}, \rho^2), (\underline{3}, \sigma\rho^2) \right\}$$

$$\mathbf{B}_{102} = \left\{ (\underline{0}, 1), (\underline{2}, 1), (\underline{0}, \sigma), (\underline{3}, \sigma), (\underline{0}, \rho), (\underline{1}, \rho), (\underline{1}, \sigma\rho), (\underline{3}, \sigma\rho) \right\}$$

$$\mathbf{B}_{103} = \left\{ (\underline{0}, 1), (\underline{2}, 1), (\underline{0}, \sigma), (\underline{3}, \sigma), (\underline{2}, \rho), (\underline{3}, \rho), (\underline{0}, \sigma\rho), (\underline{2}, \sigma\rho) \right\}$$

$$\mathbf{B}_{104} = \left\{ (\underline{0}, 1), (\underline{2}, 1), (\underline{0}, \sigma), (\underline{3}, \sigma), (\underline{0}, \rho^2), (\underline{1}, \rho^2), (\underline{2}, \rho^2), (\underline{0}, \sigma\rho^2) \right\}$$

$$\mathbf{B}_{105} = \left\{ (\underline{0}, 1), (\underline{2}, 1), (\underline{0}, \sigma), (\underline{3}, \sigma), (\underline{3}, \rho^2), (\underline{1}, \sigma\rho^2), (\underline{2}, \sigma\rho^2), (\underline{3}, \sigma\rho^2) \right\}$$

$$\mathbf{B}_{106} = \left\{ (\underline{0}, 1), (\underline{2}, 1), (\underline{1}, \sigma), (\underline{2}, \sigma), (\underline{0}, \rho), (\underline{3}, \rho), (\underline{2}, \sigma\rho), (\underline{3}, \sigma\rho) \right\}$$

$$\mathbf{B}_{107} = \left\{ (\underline{0}, 1), (\underline{2}, 1), (\underline{1}, \sigma), (\underline{2}, \sigma), (\underline{1}, \rho), (\underline{2}, \rho), (\underline{0}, \sigma\rho), (\underline{1}, \sigma\rho) \right\}$$

$$\mathbf{B}_{108} = \left\{ (\underline{0}, 1), (\underline{2}, 1), (\underline{1}, \sigma), (\underline{2}, \sigma), (\underline{0}, \rho^2), (\underline{3}, \rho^2), (\underline{0}, \sigma\rho^2), (\underline{2}, \sigma\rho^2) \right\}$$

$$\mathbf{B}_{109} = \left\{ (\underline{0}, 1), (\underline{2}, 1), (\underline{1}, \sigma), (\underline{2}, \sigma), (\underline{1}, \rho^2), (\underline{2}, \rho^2), (\underline{1}, \sigma\rho^2), (\underline{3}, \sigma\rho^2) \right\}$$

$$\mathbf{B}_{110} = \left\{ (\underline{0}, 1), (\underline{2}, 1), (\underline{1}, \sigma), (\underline{3}, \sigma), (\underline{0}, \rho), (\underline{0}, \sigma\rho), (\underline{2}, \rho^2), (\underline{2}, \sigma\rho^2) \right\}$$

$$\mathbf{B}_{111} = \left\{ (\underline{0}, 1), (\underline{2}, 1), (\underline{1}, \sigma), (\underline{3}, \sigma), (\underline{1}, \rho), (\underline{3}, \rho), (\underline{0}, \rho^2), (\underline{3}, \sigma\rho^2) \right\}$$

$$\mathbf{B}_{112} = \left\{ (\underline{0}, 1), (\underline{2}, 1), (\underline{1}, \sigma), (\underline{3}, \sigma), (\underline{2}, \rho), (\underline{3}, \sigma\rho), (\underline{1}, \rho^2), (\underline{3}, \rho^2) \right\}$$

$$\mathbf{B}_{113} = \left\{ (\underline{0}, 1), (\underline{2}, 1), (\underline{1}, \sigma), (\underline{3}, \sigma), (\underline{1}, \sigma\rho), (\underline{2}, \sigma\rho), (\underline{0}, \sigma\rho^2), (\underline{1}, \sigma\rho^2) \right\}$$

$$\mathbf{B}_{114} = \left\{ (\underline{0}, 1), (\underline{2}, 1), (\underline{2}, \sigma), (\underline{3}, \sigma), (\underline{0}, \rho), (\underline{2}, \rho), (\underline{0}, \rho^2), (\underline{1}, \sigma\rho^2) \right\}$$

$$\mathbf{B}_{115} = \left\{ (\underline{0}, 1), (\underline{2}, 1), (\underline{2}, \sigma), (\underline{3}, \sigma), (\underline{1}, \rho), (\underline{2}, \sigma\rho), (\underline{2}, \rho^2), (\underline{3}, \rho^2) \right\}$$

$$\mathbf{B_{116}} = \{ (\underline{0}, 1), (\underline{2}, 1), (\underline{2}, \sigma), (\underline{3}, \sigma), (\underline{3}, \rho), (\underline{1}, \sigma\rho), (\underline{1}, \rho^2), (\underline{2}, \sigma\rho^2) \}$$

$$\mathbf{B_{117}} = \{ (\underline{0}, 1), (\underline{2}, 1), (\underline{2}, \sigma), (\underline{3}, \sigma), (\underline{0}, \sigma\rho), (\underline{3}, \sigma\rho), (\underline{0}, \sigma\rho^2), (\underline{3}, \sigma\rho^2) \}$$

$$\mathbf{B_{118}} = \{ (\underline{0}, 1), (\underline{2}, 1), (\underline{0}, \rho), (\underline{1}, \rho), (\underline{2}, \rho), (\underline{3}, \rho), (\underline{1}, \rho^2), (\underline{2}, \rho^2) \}$$

$$\mathbf{B_{119}} = \{ (\underline{0}, 1), (\underline{2}, 1), (\underline{0}, \rho), (\underline{1}, \rho), (\underline{0}, \sigma\rho), (\underline{2}, \sigma\rho), (\underline{0}, \rho^2), (\underline{0}, \sigma\rho^2) \}$$

$$\mathbf{B_{120}} = \{ (\underline{0}, 1), (\underline{2}, 1), (\underline{0}, \rho), (\underline{2}, \rho), (\underline{1}, \sigma\rho), (\underline{2}, \sigma\rho), (\underline{3}, \rho^2), (\underline{2}, \sigma\rho^2) \}$$

$$\mathbf{B_{121}} = \{ (\underline{0}, 1), (\underline{2}, 1), (\underline{0}, \rho), (\underline{3}, \rho), (\underline{0}, \sigma\rho), (\underline{1}, \sigma\rho), (\underline{1}, \sigma\rho^2), (\underline{3}, \sigma\rho^2) \}$$

$$\mathbf{B_{122}} = \{ (\underline{0}, 1), (\underline{2}, 1), (\underline{0}, \rho), (\underline{3}, \sigma\rho), (\underline{0}, \rho^2), (\underline{2}, \rho^2), (\underline{3}, \rho^2), (\underline{3}, \sigma\rho^2) \}$$

$$\mathbf{B_{123}} = \{ (\underline{0}, 1), (\underline{2}, 1), (\underline{0}, \rho), (\underline{3}, \sigma\rho), (\underline{1}, \rho^2), (\underline{0}, \sigma\rho^2), (\underline{1}, \sigma\rho^2), (\underline{2}, \sigma\rho^2) \}$$

$$\mathbf{B_{124}} = \{ (\underline{0}, 1), (\underline{2}, 1), (\underline{1}, \rho), (\underline{2}, \rho), (\underline{2}, \sigma\rho), (\underline{3}, \sigma\rho), (\underline{1}, \sigma\rho^2), (\underline{3}, \sigma\rho^2) \}$$

$$\mathbf{B_{125}} = \{ (\underline{0}, 1), (\underline{2}, 1), (\underline{1}, \rho), (\underline{3}, \rho), (\underline{0}, \sigma\rho), (\underline{3}, \sigma\rho), (\underline{3}, \rho^2), (\underline{2}, \sigma\rho^2) \}$$

$$\mathbf{B_{126}} = \{ (\underline{0}, 1), (\underline{2}, 1), (\underline{1}, \rho), (\underline{1}, \sigma\rho), (\underline{0}, \rho^2), (\underline{2}, \rho^2), (\underline{1}, \sigma\rho^2), (\underline{2}, \sigma\rho^2) \}$$

$$\mathbf{B_{127}} = \{ (\underline{0}, 1), (\underline{2}, 1), (\underline{1}, \rho), (\underline{1}, \sigma\rho), (\underline{1}, \rho^2), (\underline{3}, \rho^2), (\underline{0}, \sigma\rho^2), (\underline{3}, \sigma\rho^2) \}$$

$$\mathbf{B_{128}} = \{ (\underline{0}, 1), (\underline{2}, 1), (\underline{2}, \rho), (\underline{3}, \rho), (\underline{1}, \sigma\rho), (\underline{3}, \sigma\rho), (\underline{0}, \rho^2), (\underline{0}, \sigma\rho^2) \}$$

$$\mathbf{B_{129}} = \{ (\underline{0}, 1), (\underline{2}, 1), (\underline{2}, \rho), (\underline{0}, \sigma\rho), (\underline{0}, \rho^2), (\underline{1}, \rho^2), (\underline{2}, \sigma\rho^2), (\underline{3}, \sigma\rho^2) \}$$

$$\mathbf{B_{130}} = \{ (\underline{0}, 1), (\underline{2}, 1), (\underline{2}, \rho), (\underline{0}, \sigma\rho), (\underline{2}, \rho^2), (\underline{3}, \rho^2), (\underline{0}, \sigma\rho^2), (\underline{1}, \sigma\rho^2) \}$$

$$\mathbf{B_{131}} = \{ (\underline{0}, 1), (\underline{2}, 1), (\underline{3}, \rho), (\underline{2}, \sigma\rho), (\underline{0}, \rho^2), (\underline{1}, \rho^2), (\underline{3}, \rho^2), (\underline{1}, \sigma\rho^2) \}$$

$$\mathbf{B_{132}} = \{ (\underline{0}, 1), (\underline{2}, 1), (\underline{3}, \rho), (\underline{2}, \sigma\rho), (\underline{2}, \rho^2), (\underline{0}, \sigma\rho^2), (\underline{2}, \sigma\rho^2), (\underline{3}, \sigma\rho^2) \}$$

$$\mathbf{B_{133}} = \{ (\underline{0}, 1), (\underline{2}, 1), (\underline{0}, \sigma\rho), (\underline{1}, \sigma\rho), (\underline{2}, \sigma\rho), (\underline{3}, \sigma\rho), (\underline{1}, \rho^2), (\underline{2}, \rho^2) \}$$

$$\mathbf{B_{134}} = \{ (\underline{0}, 1), (\underline{3}, 1), (\underline{0}, \sigma), (\underline{1}, \sigma), (\underline{0}, \rho), (\underline{3}, \sigma\rho), (\underline{1}, \rho^2), (\underline{2}, \rho^2) \}$$

$$\mathbf{B_{135}} = \{ (\underline{0}, 1), (\underline{3}, 1), (\underline{0}, \sigma), (\underline{1}, \sigma), (\underline{1}, \rho), (\underline{1}, \sigma\rho), (\underline{0}, \rho^2), (\underline{0}, \sigma\rho^2) \}$$

$$\mathbf{B_{136}} = \{ (\underline{0}, 1), (\underline{3}, 1), (\underline{0}, \sigma), (\underline{1}, \sigma), (\underline{2}, \rho), (\underline{0}, \sigma\rho), (\underline{3}, \rho^2), (\underline{2}, \sigma\rho^2) \}$$

$$\mathbf{B_{137}} = \{ (\underline{0}, 1), (\underline{3}, 1), (\underline{0}, \sigma), (\underline{1}, \sigma), (\underline{3}, \rho), (\underline{2}, \sigma\rho), (\underline{1}, \sigma\rho^2), (\underline{3}, \sigma\rho^2) \}$$

$$\mathbf{B}_{138} = \left\{ (\underline{0}, 1), (\underline{3}, 1), (\underline{0}, \sigma), (\underline{2}, \sigma), (\underline{0}, \rho), (\underline{1}, \rho), (\underline{2}, \rho), (\underline{2}, \sigma\rho) \right\}$$

$$\mathbf{B}_{139} = \left\{ (\underline{0}, 1), (\underline{3}, 1), (\underline{0}, \sigma), (\underline{2}, \sigma), (\underline{3}, \rho), (\underline{0}, \sigma\rho), (\underline{1}, \sigma\rho), (\underline{3}, \sigma\rho) \right\}$$

$$\mathbf{B}_{140} = \left\{ (\underline{0}, 1), (\underline{3}, 1), (\underline{0}, \sigma), (\underline{2}, \sigma), (\underline{0}, \rho^2), (\underline{2}, \rho^2), (\underline{3}, \rho^2), (\underline{1}, \sigma\rho^2) \right\}$$

$$\mathbf{B}_{141} = \left\{ (\underline{0}, 1), (\underline{3}, 1), (\underline{0}, \sigma), (\underline{2}, \sigma), (\underline{1}, \rho^2), (\underline{0}, \sigma\rho^2), (\underline{2}, \sigma\rho^2), (\underline{3}, \sigma\rho^2) \right\}$$

$$\mathbf{B}_{142} = \left\{ (\underline{0}, 1), (\underline{3}, 1), (\underline{0}, \sigma), (\underline{3}, \sigma), (\underline{0}, \rho), (\underline{0}, \sigma\rho), (\underline{0}, \sigma\rho^2), (\underline{1}, \sigma\rho^2) \right\}$$

$$\mathbf{B}_{143} = \left\{ (\underline{0}, 1), (\underline{3}, 1), (\underline{0}, \sigma), (\underline{3}, \sigma), (\underline{1}, \rho), (\underline{3}, \rho), (\underline{1}, \rho^2), (\underline{3}, \rho^2) \right\}$$

$$\mathbf{B}_{144} = \left\{ (\underline{0}, 1), (\underline{3}, 1), (\underline{0}, \sigma), (\underline{3}, \sigma), (\underline{2}, \rho), (\underline{3}, \sigma\rho), (\underline{0}, \rho^2), (\underline{3}, \sigma\rho^2) \right\}$$

$$\mathbf{B}_{145} = \left\{ (\underline{0}, 1), (\underline{3}, 1), (\underline{0}, \sigma), (\underline{3}, \sigma), (\underline{1}, \sigma\rho), (\underline{2}, \sigma\rho), (\underline{2}, \rho^2), (\underline{2}, \sigma\rho^2) \right\}$$

$$\mathbf{B}_{146} = \left\{ (\underline{0}, 1), (\underline{3}, 1), (\underline{1}, \sigma), (\underline{2}, \sigma), (\underline{0}, \rho), (\underline{0}, \sigma\rho), (\underline{0}, \rho^2), (\underline{3}, \sigma\rho^2) \right\}$$

$$\mathbf{B}_{147} = \left\{ (\underline{0}, 1), (\underline{3}, 1), (\underline{1}, \sigma), (\underline{2}, \sigma), (\underline{1}, \rho), (\underline{3}, \rho), (\underline{2}, \rho^2), (\underline{2}, \sigma\rho^2) \right\}$$

$$\mathbf{B}_{148} = \left\{ (\underline{0}, 1), (\underline{3}, 1), (\underline{1}, \sigma), (\underline{2}, \sigma), (\underline{2}, \rho), (\underline{3}, \sigma\rho), (\underline{0}, \sigma\rho^2), (\underline{1}, \sigma\rho^2) \right\}$$

$$\mathbf{B}_{149} = \left\{ (\underline{0}, 1), (\underline{3}, 1), (\underline{1}, \sigma), (\underline{2}, \sigma), (\underline{1}, \sigma\rho), (\underline{2}, \sigma\rho), (\underline{1}, \rho^2), (\underline{3}, \rho^2) \right\}$$

$$\mathbf{B}_{150} = \left\{ (\underline{0}, 1), (\underline{3}, 1), (\underline{1}, \sigma), (\underline{3}, \sigma), (\underline{0}, \rho), (\underline{2}, \rho), (\underline{3}, \rho), (\underline{1}, \sigma\rho) \right\}$$

$$\mathbf{B}_{151} = \left\{ (\underline{0}, 1), (\underline{3}, 1), (\underline{1}, \sigma), (\underline{3}, \sigma), (\underline{1}, \rho), (\underline{0}, \sigma\rho), (\underline{2}, \sigma\rho), (\underline{3}, \sigma\rho) \right\}$$

$$\mathbf{B}_{152} = \left\{ (\underline{0}, 1), (\underline{3}, 1), (\underline{1}, \sigma), (\underline{3}, \sigma), (\underline{0}, \rho^2), (\underline{1}, \rho^2), (\underline{1}, \sigma\rho^2), (\underline{2}, \sigma\rho^2) \right\}$$

$$\mathbf{B}_{153} = \left\{ (\underline{0}, 1), (\underline{3}, 1), (\underline{1}, \sigma), (\underline{3}, \sigma), (\underline{2}, \rho^2), (\underline{3}, \rho^2), (\underline{0}, \sigma\rho^2), (\underline{3}, \sigma\rho^2) \right\}$$

$$\mathbf{B}_{154} = \left\{ (\underline{0}, 1), (\underline{3}, 1), (\underline{2}, \sigma), (\underline{3}, \sigma), (\underline{0}, \rho), (\underline{3}, \sigma\rho), (\underline{3}, \rho^2), (\underline{2}, \sigma\rho^2) \right\}$$

$$\mathbf{B}_{155} = \left\{ (\underline{0}, 1), (\underline{3}, 1), (\underline{2}, \sigma), (\underline{3}, \sigma), (\underline{1}, \rho), (\underline{1}, \sigma\rho), (\underline{1}, \sigma\rho^2), (\underline{3}, \sigma\rho^2) \right\}$$

$$\mathbf{B}_{156} = \left\{ (\underline{0}, 1), (\underline{3}, 1), (\underline{2}, \sigma), (\underline{3}, \sigma), (\underline{2}, \rho), (\underline{0}, \sigma\rho), (\underline{1}, \rho^2), (\underline{2}, \rho^2) \right\}$$

$$\mathbf{B}_{157} = \left\{ (\underline{0}, 1), (\underline{3}, 1), (\underline{2}, \sigma), (\underline{3}, \sigma), (\underline{3}, \rho), (\underline{2}, \sigma\rho), (\underline{0}, \rho^2), (\underline{0}, \sigma\rho^2) \right\}$$

$$\mathbf{B}_{158} = \left\{ (\underline{0}, 1), (\underline{3}, 1), (\underline{0}, \rho), (\underline{1}, \rho), (\underline{3}, \rho), (\underline{3}, \sigma\rho), (\underline{0}, \sigma\rho^2), (\underline{3}, \sigma\rho^2) \right\}$$

$$\mathbf{B}_{159} = \left\{ (\underline{0}, 1), (\underline{3}, 1), (\underline{0}, \rho), (\underline{1}, \rho), (\underline{0}, \sigma\rho), (\underline{1}, \sigma\rho), (\underline{1}, \rho^2), (\underline{2}, \sigma\rho^2) \right\}$$

$$\mathbf{B}_{160} = \left\{ (\underline{0}, 1), (\underline{3}, 1), (\underline{0}, \rho), (\underline{2}, \rho), (\underline{0}, \rho^2), (\underline{1}, \rho^2), (\underline{3}, \rho^2), (\underline{0}, \sigma\rho^2) \right\}$$

$$\mathbf{B}_{161} = \left\{ (\underline{0}, 1), (\underline{3}, 1), (\underline{0}, \rho), (\underline{2}, \rho), (\underline{2}, \rho^2), (\underline{1}, \sigma\rho^2), (\underline{2}, \sigma\rho^2), (\underline{3}, \sigma\rho^2) \right\}$$

$$\mathbf{B}_{162} = \left\{ (\underline{0}, 1), (\underline{3}, 1), (\underline{0}, \rho), (\underline{3}, \rho), (\underline{0}, \sigma\rho), (\underline{2}, \sigma\rho), (\underline{2}, \rho^2), (\underline{3}, \rho^2) \right\}$$

$$\mathbf{B}_{163} = \left\{ (\underline{0}, 1), (\underline{3}, 1), (\underline{0}, \rho), (\underline{1}, \sigma\rho), (\underline{2}, \sigma\rho), (\underline{3}, \sigma\rho), (\underline{0}, \rho^2), (\underline{1}, \sigma\rho^2) \right\}$$

$$\mathbf{B}_{164} = \left\{ (\underline{0}, 1), (\underline{3}, 1), (\underline{1}, \rho), (\underline{2}, \rho), (\underline{3}, \rho), (\underline{0}, \sigma\rho), (\underline{0}, \rho^2), (\underline{1}, \sigma\rho^2) \right\}$$

$$\mathbf{B}_{165} = \left\{ (\underline{0}, 1), (\underline{3}, 1), (\underline{1}, \rho), (\underline{2}, \rho), (\underline{1}, \sigma\rho), (\underline{3}, \sigma\rho), (\underline{2}, \rho^2), (\underline{3}, \rho^2) \right\}$$

$$\mathbf{B}_{166} = \left\{ (\underline{0}, 1), (\underline{3}, 1), (\underline{1}, \rho), (\underline{2}, \sigma\rho), (\underline{0}, \rho^2), (\underline{3}, \rho^2), (\underline{2}, \sigma\rho^2), (\underline{3}, \sigma\rho^2) \right\}$$

$$\mathbf{B}_{167} = \left\{ (\underline{0}, 1), (\underline{3}, 1), (\underline{1}, \rho), (\underline{2}, \sigma\rho), (\underline{1}, \rho^2), (\underline{2}, \rho^2), (\underline{0}, \sigma\rho^2), (\underline{1}, \sigma\rho^2) \right\}$$

$$\mathbf{B}_{168} = \left\{ (\underline{0}, 1), (\underline{3}, 1), (\underline{2}, \rho), (\underline{3}, \rho), (\underline{2}, \sigma\rho), (\underline{3}, \sigma\rho), (\underline{1}, \rho^2), (\underline{2}, \sigma\rho^2) \right\}$$

$$\mathbf{B}_{169} = \left\{ (\underline{0}, 1), (\underline{3}, 1), (\underline{2}, \rho), (\underline{0}, \sigma\rho), (\underline{1}, \sigma\rho), (\underline{2}, \sigma\rho), (\underline{0}, \sigma\rho^2), (\underline{3}, \sigma\rho^2) \right\}$$

$$\mathbf{B}_{170} = \left\{ (\underline{0}, 1), (\underline{3}, 1), (\underline{3}, \rho), (\underline{1}, \sigma\rho), (\underline{0}, \rho^2), (\underline{1}, \rho^2), (\underline{2}, \rho^2), (\underline{3}, \sigma\rho^2) \right\}$$

$$\mathbf{B}_{171} = \left\{ (\underline{0}, 1), (\underline{3}, 1), (\underline{3}, \rho), (\underline{1}, \sigma\rho), (\underline{3}, \rho^2), (\underline{0}, \sigma\rho^2), (\underline{1}, \sigma\rho^2), (\underline{2}, \sigma\rho^2) \right\}$$

$$\mathbf{B}_{172} = \left\{ (\underline{0}, 1), (\underline{3}, 1), (\underline{0}, \sigma\rho), (\underline{3}, \sigma\rho), (\underline{0}, \rho^2), (\underline{2}, \rho^2), (\underline{0}, \sigma\rho^2), (\underline{2}, \sigma\rho^2) \right\}$$

$$\mathbf{B}_{173} = \left\{ (\underline{0}, 1), (\underline{3}, 1), (\underline{0}, \sigma\rho), (\underline{3}, \sigma\rho), (\underline{1}, \rho^2), (\underline{3}, \rho^2), (\underline{1}, \sigma\rho^2), (\underline{3}, \sigma\rho^2) \right\}$$

$$\mathbf{B}_{174} = \left\{ (\underline{0}, 1), (\underline{0}, \sigma), (\underline{1}, \sigma), (\underline{2}, \sigma), (\underline{0}, \rho), (\underline{1}, \sigma\rho), (\underline{1}, \sigma\rho^2), (\underline{2}, \sigma\rho^2) \right\}$$

$$\mathbf{B}_{175} = \left\{ (\underline{0}, 1), (\underline{0}, \sigma), (\underline{1}, \sigma), (\underline{2}, \sigma), (\underline{1}, \rho), (\underline{3}, \sigma\rho), (\underline{3}, \rho^2), (\underline{3}, \sigma\rho^2) \right\}$$

$$\mathbf{B}_{176} = \left\{ (\underline{0}, 1), (\underline{0}, \sigma), (\underline{1}, \sigma), (\underline{2}, \sigma), (\underline{2}, \rho), (\underline{3}, \rho), (\underline{0}, \rho^2), (\underline{1}, \rho^2) \right\}$$

$$\mathbf{B}_{177} = \left\{ (\underline{0}, 1), (\underline{0}, \sigma), (\underline{1}, \sigma), (\underline{2}, \sigma), (\underline{0}, \sigma\rho), (\underline{2}, \sigma\rho), (\underline{2}, \rho^2), (\underline{0}, \sigma\rho^2) \right\}$$

$$\mathbf{B}_{178} = \left\{ (\underline{0}, 1), (\underline{0}, \sigma), (\underline{1}, \sigma), (\underline{3}, \sigma), (\underline{0}, \rho), (\underline{2}, \sigma\rho), (\underline{0}, \rho^2), (\underline{3}, \rho^2) \right\}$$

$$\mathbf{B}_{179} = \left\{ (\underline{0}, 1), (\underline{0}, \sigma), (\underline{1}, \sigma), (\underline{3}, \sigma), (\underline{1}, \rho), (\underline{2}, \rho), (\underline{2}, \rho^2), (\underline{1}, \sigma\rho^2) \right\}$$

$$\mathbf{B}_{180} = \left\{ (\underline{0}, 1), (\underline{0}, \sigma), (\underline{1}, \sigma), (\underline{3}, \sigma), (\underline{3}, \rho), (\underline{3}, \sigma\rho), (\underline{0}, \sigma\rho^2), (\underline{2}, \sigma\rho^2) \right\}$$

$$\mathbf{B}_{181} = \left\{ (\underline{0}, 1), (\underline{0}, \sigma), (\underline{1}, \sigma), (\underline{3}, \sigma), (\underline{0}, \sigma\rho), (\underline{1}, \sigma\rho), (\underline{1}, \rho^2), (\underline{3}, \sigma\rho^2) \right\}$$

$$\mathbf{B}_{182} = \left\{ (\underline{0}, 1), (\underline{0}, \sigma), (\underline{2}, \sigma), (\underline{3}, \sigma), (\underline{0}, \rho), (\underline{3}, \rho), (\underline{2}, \rho^2), (\underline{3}, \sigma\rho^2) \right\}$$

$$\mathbf{B}_{183} = \left\{ (\underline{0}, 1), (\underline{0}, \sigma), (\underline{2}, \sigma), (\underline{3}, \sigma), (\underline{1}, \rho), (\underline{0}, \sigma\rho), (\underline{0}, \rho^2), (\underline{2}, \sigma\rho^2) \right\}$$

$$\mathbf{B}_{184} = \left\{ (\underline{0}, 1), (\underline{0}, \sigma), (\underline{2}, \sigma), (\underline{3}, \sigma), (\underline{2}, \rho), (\underline{1}, \sigma\rho), (\underline{3}, \rho^2), (\underline{0}, \sigma\rho^2) \right\}$$

$$\mathbf{B}_{185} = \left\{ (\underline{0}, 1), (\underline{0}, \sigma), (\underline{2}, \sigma), (\underline{3}, \sigma), (\underline{2}, \sigma\rho), (\underline{3}, \sigma\rho), (\underline{1}, \rho^2), (\underline{1}, \sigma\rho^2) \right\}$$

$$\mathbf{B}_{186} = \left\{ (\underline{0}, 1), (\underline{0}, \sigma), (\underline{0}, \rho), (\underline{1}, \rho), (\underline{0}, \rho^2), (\underline{1}, \rho^2), (\underline{1}, \sigma\rho^2), (\underline{3}, \sigma\rho^2) \right\}$$

$$\mathbf{B}_{187} = \left\{ (\underline{0}, 1), (\underline{0}, \sigma), (\underline{0}, \rho), (\underline{1}, \rho), (\underline{2}, \rho^2), (\underline{3}, \rho^2), (\underline{0}, \sigma\rho^2), (\underline{2}, \sigma\rho^2) \right\}$$

$$\mathbf{B}_{188} = \left\{ (\underline{0}, 1), (\underline{0}, \sigma), (\underline{0}, \rho), (\underline{2}, \rho), (\underline{3}, \rho), (\underline{3}, \sigma\rho), (\underline{3}, \rho^2), (\underline{1}, \sigma\rho^2) \right\}$$

$$\mathbf{B}_{189} = \left\{ (\underline{0}, 1), (\underline{0}, \sigma), (\underline{0}, \rho), (\underline{2}, \rho), (\underline{0}, \sigma\rho), (\underline{1}, \sigma\rho), (\underline{0}, \rho^2), (\underline{2}, \rho^2) \right\}$$

$$\mathbf{B}_{190} = \left\{ (\underline{0}, 1), (\underline{0}, \sigma), (\underline{0}, \rho), (\underline{3}, \rho), (\underline{1}, \sigma\rho), (\underline{2}, \sigma\rho), (\underline{1}, \rho^2), (\underline{0}, \sigma\rho^2) \right\}$$

$$\mathbf{B}_{191} = \left\{ (\underline{0}, 1), (\underline{0}, \sigma), (\underline{0}, \rho), (\underline{0}, \sigma\rho), (\underline{2}, \sigma\rho), (\underline{3}, \sigma\rho), (\underline{2}, \sigma\rho^2), (\underline{3}, \sigma\rho^2) \right\}$$

$$\mathbf{B}_{192} = \left\{ (\underline{0}, 1), (\underline{0}, \sigma), (\underline{1}, \rho), (\underline{2}, \rho), (\underline{3}, \rho), (\underline{1}, \sigma\rho), (\underline{2}, \sigma\rho^2), (\underline{3}, \sigma\rho^2) \right\}$$

$$\mathbf{B}_{193} = \left\{ (\underline{0}, 1), (\underline{0}, \sigma), (\underline{1}, \rho), (\underline{2}, \rho), (\underline{0}, \sigma\rho), (\underline{3}, \sigma\rho), (\underline{1}, \rho^2), (\underline{0}, \sigma\rho^2) \right\}$$

$$\mathbf{B}_{194} = \left\{ (\underline{0}, 1), (\underline{0}, \sigma), (\underline{1}, \rho), (\underline{3}, \rho), (\underline{2}, \sigma\rho), (\underline{3}, \sigma\rho), (\underline{0}, \rho^2), (\underline{2}, \rho^2) \right\}$$

$$\mathbf{B}_{195} = \left\{ (\underline{0}, 1), (\underline{0}, \sigma), (\underline{1}, \rho), (\underline{0}, \sigma\rho), (\underline{1}, \sigma\rho), (\underline{2}, \sigma\rho), (\underline{3}, \rho^2), (\underline{1}, \sigma\rho^2) \right\}$$

$$\mathbf{B}_{196} = \left\{ (\underline{0}, 1), (\underline{0}, \sigma), (\underline{2}, \rho), (\underline{2}, \sigma\rho), (\underline{0}, \rho^2), (\underline{0}, \sigma\rho^2), (\underline{1}, \sigma\rho^2), (\underline{2}, \sigma\rho^2) \right\}$$

$$\mathbf{B}_{197} = \left\{ (\underline{0}, 1), (\underline{0}, \sigma), (\underline{2}, \rho), (\underline{2}, \sigma\rho), (\underline{1}, \rho^2), (\underline{2}, \rho^2), (\underline{3}, \rho^2), (\underline{3}, \sigma\rho^2) \right\}$$

$$\mathbf{B}_{198} = \left\{ (\underline{0}, 1), (\underline{0}, \sigma), (\underline{3}, \rho), (\underline{0}, \sigma\rho), (\underline{0}, \rho^2), (\underline{3}, \rho^2), (\underline{0}, \sigma\rho^2), (\underline{3}, \sigma\rho^2) \right\}$$

$$\mathbf{B}_{199} = \left\{ (\underline{0}, 1), (\underline{0}, \sigma), (\underline{3}, \rho), (\underline{0}, \sigma\rho), (\underline{1}, \rho^2), (\underline{2}, \rho^2), (\underline{1}, \sigma\rho^2), (\underline{2}, \sigma\rho^2) \right\}$$

$$\mathbf{B}_{200} = \left\{ (\underline{0}, 1), (\underline{0}, \sigma), (\underline{1}, \sigma\rho), (\underline{3}, \sigma\rho), (\underline{0}, \rho^2), (\underline{1}, \rho^2), (\underline{3}, \rho^2), (\underline{2}, \sigma\rho^2) \right\}$$

$$\mathbf{B}_{201} = \left\{ (\underline{0}, 1), (\underline{0}, \sigma), (\underline{1}, \sigma\rho), (\underline{3}, \sigma\rho), (\underline{2}, \rho^2), (\underline{0}, \sigma\rho^2), (\underline{1}, \sigma\rho^2), (\underline{3}, \sigma\rho^2) \right\}$$

$$\mathbf{B}_{202} = \left\{ (\underline{0}, 1), (\underline{1}, \sigma), (\underline{2}, \sigma), (\underline{3}, \sigma), (\underline{0}, \rho), (\underline{1}, \rho), (\underline{1}, \rho^2), (\underline{0}, \sigma\rho^2) \right\}$$

$$\mathbf{B}_{203} = \left\{ (\underline{0}, 1), (\underline{1}, \sigma), (\underline{2}, \sigma), (\underline{3}, \sigma), (\underline{2}, \rho), (\underline{2}, \sigma\rho), (\underline{2}, \sigma\rho^2), (\underline{3}, \sigma\rho^2) \right\}$$

$$\mathbf{B}_{204} = \left\{ (\underline{0}, 1), (\underline{1}, \sigma), (\underline{2}, \sigma), (\underline{3}, \sigma), (\underline{3}, \rho), (\underline{0}, \sigma\rho), (\underline{3}, \rho^2), (\underline{1}, \sigma\rho^2) \right\}$$

$$\mathbf{B}_{205} = \left\{ (\underline{0}, 1), (\underline{1}, \sigma), (\underline{2}, \sigma), (\underline{3}, \sigma), (\underline{1}, \sigma\rho), (\underline{3}, \sigma\rho), (\underline{0}, \rho^2), (\underline{2}, \rho^2) \right\}$$

$$\mathbf{B}_{206} = \left\{ (\underline{0}, 1), (\underline{1}, \sigma), (\underline{0}, \rho), (\underline{1}, \rho), (\underline{2}, \rho), (\underline{3}, \sigma\rho), (\underline{0}, \rho^2), (\underline{2}, \sigma\rho^2) \right\}$$

$$\mathbf{B}_{207} = \left\{ (\underline{0}, 1), (\underline{1}, \sigma), (\underline{0}, \rho), (\underline{1}, \rho), (\underline{1}, \sigma\rho), (\underline{2}, \sigma\rho), (\underline{2}, \rho^2), (\underline{3}, \sigma\rho^2) \right\}$$

$$\mathbf{B}_{208} = \left\{ (\underline{0}, 1), (\underline{1}, \sigma), (\underline{0}, \rho), (\underline{2}, \rho), (\underline{0}, \sigma\rho), (\underline{2}, \sigma\rho), (\underline{1}, \rho^2), (\underline{1}, \sigma\rho^2) \right\}$$

$$\mathbf{B}_{209} = \left\{ (\underline{0}, 1), (\underline{1}, \sigma), (\underline{0}, \rho), (\underline{3}, \rho), (\underline{0}, \rho^2), (\underline{2}, \rho^2), (\underline{0}, \sigma\rho^2), (\underline{1}, \sigma\rho^2) \right\}$$

$$\mathbf{B}_{210} = \left\{ (\underline{0}, 1), (\underline{1}, \sigma), (\underline{0}, \rho), (\underline{3}, \rho), (\underline{1}, \rho^2), (\underline{3}, \rho^2), (\underline{2}, \sigma\rho^2), (\underline{3}, \sigma\rho^2) \right\}$$

$$\mathbf{B}_{211} = \left\{ (\underline{0}, 1), (\underline{1}, \sigma), (\underline{0}, \rho), (\underline{0}, \sigma\rho), (\underline{1}, \sigma\rho), (\underline{3}, \sigma\rho), (\underline{3}, \rho^2), (\underline{0}, \sigma\rho^2) \right\}$$

$$\mathbf{B}_{212} = \left\{ (\underline{0}, 1), (\underline{1}, \sigma), (\underline{1}, \rho), (\underline{2}, \rho), (\underline{3}, \rho), (\underline{2}, \sigma\rho), (\underline{3}, \rho^2), (\underline{0}, \sigma\rho^2) \right\}$$

$$\mathbf{B}_{213} = \left\{ (\underline{0}, 1), (\underline{1}, \sigma), (\underline{1}, \rho), (\underline{3}, \rho), (\underline{1}, \sigma\rho), (\underline{3}, \sigma\rho), (\underline{1}, \rho^2), (\underline{1}, \sigma\rho^2) \right\}$$

$$\mathbf{B}_{214} = \left\{ (\underline{0}, 1), (\underline{1}, \sigma), (\underline{1}, \rho), (\underline{0}, \sigma\rho), (\underline{0}, \rho^2), (\underline{1}, \rho^2), (\underline{2}, \rho^2), (\underline{3}, \rho^2) \right\}$$

$$\mathbf{B}_{215} = \left\{ (\underline{0}, 1), (\underline{1}, \sigma), (\underline{1}, \rho), (\underline{0}, \sigma\rho), (\underline{0}, \sigma\rho^2), (\underline{1}, \sigma\rho^2), (\underline{2}, \sigma\rho^2), (\underline{3}, \sigma\rho^2) \right\}$$

$$\mathbf{B}_{216} = \left\{ (\underline{0}, 1), (\underline{1}, \sigma), (\underline{2}, \rho), (\underline{3}, \rho), (\underline{0}, \sigma\rho), (\underline{3}, \sigma\rho), (\underline{2}, \rho^2), (\underline{3}, \sigma\rho^2) \right\}$$

$$\mathbf{B}_{217} = \left\{ (\underline{0}, 1), (\underline{1}, \sigma), (\underline{2}, \rho), (\underline{1}, \sigma\rho), (\underline{0}, \rho^2), (\underline{3}, \rho^2), (\underline{1}, \sigma\rho^2), (\underline{3}, \sigma\rho^2) \right\}$$

$$\mathbf{B}_{218} = \left\{ (\underline{0}, 1), (\underline{1}, \sigma), (\underline{2}, \rho), (\underline{1}, \sigma\rho), (\underline{1}, \rho^2), (\underline{2}, \rho^2), (\underline{0}, \sigma\rho^2), (\underline{2}, \sigma\rho^2) \right\}$$

$$\mathbf{B}_{219} = \left\{ (\underline{0}, 1), (\underline{1}, \sigma), (\underline{3}, \rho), (\underline{0}, \sigma\rho), (\underline{1}, \sigma\rho), (\underline{2}, \sigma\rho), (\underline{0}, \rho^2), (\underline{2}, \sigma\rho^2) \right\}$$

$$\mathbf{B}_{220} = \left\{ (\underline{0}, 1), (\underline{1}, \sigma), (\underline{2}, \sigma\rho), (\underline{3}, \sigma\rho), (\underline{0}, \rho^2), (\underline{1}, \rho^2), (\underline{0}, \sigma\rho^2), (\underline{3}, \sigma\rho^2) \right\}$$

$$\mathbf{B}_{221} = \left\{ (\underline{0}, 1), (\underline{1}, \sigma), (\underline{2}, \sigma\rho), (\underline{3}, \sigma\rho), (\underline{2}, \rho^2), (\underline{3}, \rho^2), (\underline{1}, \sigma\rho^2), (\underline{2}, \sigma\rho^2) \right\}$$

$$\mathbf{B}_{222} = \left\{ (\underline{0}, 1), (\underline{2}, \sigma), (\underline{0}, \rho), (\underline{1}, \rho), (\underline{3}, \rho), (\underline{1}, \sigma\rho), (\underline{0}, \rho^2), (\underline{3}, \rho^2) \right\}$$

$$\mathbf{B}_{223} = \left\{ (\underline{0}, 1), (\underline{2}, \sigma), (\underline{0}, \rho), (\underline{1}, \rho), (\underline{0}, \sigma\rho), (\underline{3}, \sigma\rho), (\underline{2}, \rho^2), (\underline{1}, \sigma\rho^2) \right\}$$

$$\mathbf{B}_{224} = \left\{ (\underline{0}, 1), (\underline{2}, \sigma), (\underline{0}, \rho), (\underline{2}, \rho), (\underline{3}, \rho), (\underline{0}, \sigma\rho), (\underline{0}, \sigma\rho^2), (\underline{2}, \sigma\rho^2) \right\}$$

$$\mathbf{B}_{225} = \left\{ (\underline{0}, 1), (\underline{2}, \sigma), (\underline{0}, \rho), (\underline{2}, \rho), (\underline{1}, \sigma\rho), (\underline{3}, \sigma\rho), (\underline{1}, \rho^2), (\underline{3}, \sigma\rho^2) \right\}$$

$$\mathbf{B}_{226} = \left\{ (\underline{0}, 1), (\underline{2}, \sigma), (\underline{0}, \rho), (\underline{2}, \sigma\rho), (\underline{0}, \rho^2), (\underline{1}, \rho^2), (\underline{2}, \rho^2), (\underline{2}, \sigma\rho^2) \right\}$$

$$\mathbf{B}_{227} = \left\{ (\underline{0}, 1), (\underline{2}, \sigma), (\underline{0}, \rho), (\underline{2}, \sigma\rho), (\underline{3}, \rho^2), (\underline{0}, \sigma\rho^2), (\underline{1}, \sigma\rho^2), (\underline{3}, \sigma\rho^2) \right\}$$

$$\mathbf{B}_{228} = \left\{ (\underline{0}, 1), (\underline{2}, \sigma), (\underline{1}, \rho), (\underline{2}, \rho), (\underline{0}, \rho^2), (\underline{2}, \rho^2), (\underline{0}, \sigma\rho^2), (\underline{3}, \sigma\rho^2) \right\}$$

$$\mathbf{B}_{229} = \left\{ (\underline{0}, 1), (\underline{2}, \sigma), (\underline{1}, \rho), (\underline{2}, \rho), (\underline{1}, \rho^2), (\underline{3}, \rho^2), (\underline{1}, \sigma\rho^2), (\underline{2}, \sigma\rho^2) \right\}$$

$$\mathbf{B}_{230} = \left\{ (\underline{0}, 1), (\underline{2}, \sigma), (\underline{1}, \rho), (\underline{3}, \rho), (\underline{0}, \sigma\rho), (\underline{2}, \sigma\rho), (\underline{1}, \rho^2), (\underline{3}, \sigma\rho^2) \right\}$$

$$\mathbf{B}_{231} = \left\{ (\underline{0}, 1), (\underline{2}, \sigma), (\underline{1}, \rho), (\underline{1}, \sigma\rho), (\underline{2}, \sigma\rho), (\underline{3}, \sigma\rho), (\underline{0}, \sigma\rho^2), (\underline{2}, \sigma\rho^2) \right\}$$

$$\mathbf{B}_{232} = \left\{ (\underline{0}, 1), (\underline{2}, \sigma), (\underline{2}, \rho), (\underline{3}, \rho), (\underline{1}, \sigma\rho), (\underline{2}, \sigma\rho), (\underline{2}, \rho^2), (\underline{1}, \sigma\rho^2) \right\}$$

$$\mathbf{B}_{233} = \left\{ (\underline{0}, 1), (\underline{2}, \sigma), (\underline{2}, \rho), (\underline{0}, \sigma\rho), (\underline{2}, \sigma\rho), (\underline{3}, \sigma\rho), (\underline{0}, \rho^2), (\underline{3}, \rho^2) \right\}$$

$$\mathbf{B}_{234} = \left\{ (\underline{0}, 1), (\underline{2}, \sigma), (\underline{3}, \rho), (\underline{3}, \sigma\rho), (\underline{0}, \rho^2), (\underline{1}, \sigma\rho^2), (\underline{2}, \sigma\rho^2), (\underline{3}, \sigma\rho^2) \right\}$$

$$\mathbf{B}_{235} = \left\{ (\underline{0}, 1), (\underline{2}, \sigma), (\underline{3}, \rho), (\underline{3}, \sigma\rho), (\underline{1}, \rho^2), (\underline{2}, \rho^2), (\underline{3}, \rho^2), (\underline{0}, \sigma\rho^2) \right\}$$

$$\mathbf{B}_{236} = \left\{ (\underline{0}, 1), (\underline{2}, \sigma), (\underline{0}, \sigma\rho), (\underline{1}, \sigma\rho), (\underline{0}, \rho^2), (\underline{1}, \rho^2), (\underline{0}, \sigma\rho^2), (\underline{1}, \sigma\rho^2) \right\}$$

$$\mathbf{B}_{237} = \left\{ (\underline{0}, 1), (\underline{2}, \sigma), (\underline{0}, \sigma\rho), (\underline{1}, \sigma\rho), (\underline{2}, \rho^2), (\underline{3}, \rho^2), (\underline{2}, \sigma\rho^2), (\underline{3}, \sigma\rho^2) \right\}$$

$$\mathbf{B}_{238} = \left\{ (\underline{0}, 1), (\underline{3}, \sigma), (\underline{0}, \rho), (\underline{1}, \rho), (\underline{2}, \rho), (\underline{0}, \sigma\rho), (\underline{3}, \rho^2), (\underline{3}, \sigma\rho^2) \right\}$$

$$\mathbf{B}_{239} = \left\{ (\underline{0}, 1), (\underline{3}, \sigma), (\underline{0}, \rho), (\underline{1}, \rho), (\underline{3}, \rho), (\underline{2}, \sigma\rho), (\underline{1}, \sigma\rho^2), (\underline{2}, \sigma\rho^2) \right\}$$

$$\mathbf{B}_{240} = \left\{ (\underline{0}, 1), (\underline{3}, \sigma), (\underline{0}, \rho), (\underline{2}, \rho), (\underline{2}, \sigma\rho), (\underline{3}, \sigma\rho), (\underline{2}, \rho^2), (\underline{0}, \sigma\rho^2) \right\}$$

$$\mathbf{B}_{241} = \left\{ (\underline{0}, 1), (\underline{3}, \sigma), (\underline{0}, \rho), (\underline{3}, \rho), (\underline{0}, \sigma\rho), (\underline{3}, \sigma\rho), (\underline{0}, \rho^2), (\underline{1}, \rho^2) \right\}$$

$$\mathbf{B}_{242} = \left\{ (\underline{0}, 1), (\underline{3}, \sigma), (\underline{0}, \rho), (\underline{1}, \sigma\rho), (\underline{0}, \rho^2), (\underline{0}, \sigma\rho^2), (\underline{2}, \sigma\rho^2), (\underline{3}, \sigma\rho^2) \right\}$$

$$\mathbf{B}_{243} = \left\{ (\underline{0}, 1), (\underline{3}, \sigma), (\underline{0}, \rho), (\underline{1}, \sigma\rho), (\underline{1}, \rho^2), (\underline{2}, \rho^2), (\underline{3}, \rho^2), (\underline{1}, \sigma\rho^2) \right\}$$

$$\mathbf{B}_{244} = \left\{ (\underline{0}, 1), (\underline{3}, \sigma), (\underline{1}, \rho), (\underline{2}, \rho), (\underline{1}, \sigma\rho), (\underline{2}, \sigma\rho), (\underline{0}, \rho^2), (\underline{1}, \rho^2) \right\}$$

$$\mathbf{B}_{245} = \left\{ (\underline{0}, 1), (\underline{3}, \sigma), (\underline{1}, \rho), (\underline{3}, \rho), (\underline{0}, \sigma\rho), (\underline{1}, \sigma\rho), (\underline{2}, \rho^2), (\underline{0}, \sigma\rho^2) \right\}$$

$$\mathbf{B}_{246} = \left\{ (\underline{0}, 1), (\underline{3}, \sigma), (\underline{1}, \rho), (\underline{3}, \sigma\rho), (\underline{0}, \rho^2), (\underline{3}, \rho^2), (\underline{0}, \sigma\rho^2), (\underline{1}, \sigma\rho^2) \right\}$$

$$\mathbf{B}_{247} = \left\{ (\underline{0}, 1), (\underline{3}, \sigma), (\underline{1}, \rho), (\underline{3}, \sigma\rho), (\underline{1}, \rho^2), (\underline{2}, \rho^2), (\underline{2}, \sigma\rho^2), (\underline{3}, \sigma\rho^2) \right\}$$

$$\mathbf{B}_{248} = \left\{ (\underline{0}, 1), (\underline{3}, \sigma), (\underline{2}, \rho), (\underline{3}, \rho), (\underline{0}, \rho^2), (\underline{2}, \rho^2), (\underline{3}, \rho^2), (\underline{2}, \sigma\rho^2) \right\}$$

$$\mathbf{B}_{249} = \left\{ (\underline{0}, 1), (\underline{3}, \sigma), (\underline{2}, \rho), (\underline{3}, \rho), (\underline{1}, \rho^2), (\underline{0}, \sigma\rho^2), (\underline{1}, \sigma\rho^2), (\underline{3}, \sigma\rho^2) \right\}$$

$$\mathbf{B}_{250} = \left\{ (\underline{0}, 1), (\underline{3}, \sigma), (\underline{2}, \rho), (\underline{0}, \sigma\rho), (\underline{1}, \sigma\rho), (\underline{3}, \sigma\rho), (\underline{1}, \sigma\rho^2), (\underline{2}, \sigma\rho^2) \right\}$$

$$\mathbf{B}_{251} = \left\{ (\underline{0}, 1), (\underline{3}, \sigma), (\underline{3}, \rho), (\underline{1}, \sigma\rho), (\underline{2}, \sigma\rho), (\underline{3}, \sigma\rho), (\underline{3}, \rho^2), (\underline{3}, \sigma\rho^2) \right\}$$

$$\mathbf{B}_{252} = \left\{ (\underline{0}, 1), (\underline{3}, \sigma), (\underline{0}, \sigma\rho), (\underline{2}, \sigma\rho), (\underline{0}, \rho^2), (\underline{2}, \rho^2), (\underline{1}, \sigma\rho^2), (\underline{3}, \sigma\rho^2) \right\}$$

$$\mathbf{B}_{253} = \left\{ (\underline{0}, 1), (\underline{3}, \sigma), (\underline{0}, \sigma\rho), (\underline{2}, \sigma\rho), (\underline{1}, \rho^2), (\underline{3}, \rho^2), (\underline{0}, \sigma\rho^2), (\underline{2}, \sigma\rho^2) \right\}$$

$$\mathbf{B}_{254} = \left\{ (\underline{1}, 1), (\underline{2}, 1), (\underline{3}, 1), (\underline{0}, \sigma), (\underline{0}, \rho), (\underline{1}, \rho), (\underline{1}, \rho^2), (\underline{0}, \sigma\rho^2) \right\}$$

$$\mathbf{B}_{255} = \left\{ (\underline{1}, 1), (\underline{2}, 1), (\underline{3}, 1), (\underline{0}, \sigma), (\underline{2}, \rho), (\underline{2}, \sigma\rho), (\underline{2}, \sigma\rho^2), (\underline{3}, \sigma\rho^2) \right\}$$

$$\mathbf{B}_{256} = \left\{ (\underline{1}, 1), (\underline{2}, 1), (\underline{3}, 1), (\underline{0}, \sigma), (\underline{3}, \rho), (\underline{0}, \sigma\rho), (\underline{3}, \rho^2), (\underline{1}, \sigma\rho^2) \right\}$$

$$\mathbf{B}_{257} = \left\{ (\underline{1}, 1), (\underline{2}, 1), (\underline{3}, 1), (\underline{0}, \sigma), (\underline{1}, \sigma\rho), (\underline{3}, \sigma\rho), (\underline{0}, \rho^2), (\underline{2}, \rho^2) \right\}$$

$$\mathbf{B}_{258} = \left\{ (\underline{1}, 1), (\underline{2}, 1), (\underline{3}, 1), (\underline{1}, \sigma), (\underline{0}, \rho), (\underline{3}, \rho), (\underline{2}, \rho^2), (\underline{3}, \sigma\rho^2) \right\}$$

$$\mathbf{B}_{259} = \left\{ (\underline{1}, 1), (\underline{2}, 1), (\underline{3}, 1), (\underline{1}, \sigma), (\underline{1}, \rho), (\underline{0}, \sigma\rho), (\underline{0}, \rho^2), (\underline{2}, \sigma\rho^2) \right\}$$

$$\mathbf{B}_{260} = \left\{ (\underline{1}, 1), (\underline{2}, 1), (\underline{3}, 1), (\underline{1}, \sigma), (\underline{2}, \rho), (\underline{1}, \sigma\rho), (\underline{3}, \rho^2), (\underline{0}, \sigma\rho^2) \right\}$$

$$\mathbf{B}_{261} = \left\{ (\underline{1}, 1), (\underline{2}, 1), (\underline{3}, 1), (\underline{1}, \sigma), (\underline{2}, \sigma\rho), (\underline{3}, \sigma\rho), (\underline{1}, \rho^2), (\underline{1}, \sigma\rho^2) \right\}$$

$$\mathbf{B}_{262} = \left\{ (\underline{1}, 1), (\underline{2}, 1), (\underline{3}, 1), (\underline{2}, \sigma), (\underline{0}, \rho), (\underline{2}, \sigma\rho), (\underline{0}, \rho^2), (\underline{3}, \rho^2) \right\}$$

$$\mathbf{B}_{263} = \left\{ (\underline{1}, 1), (\underline{2}, 1), (\underline{3}, 1), (\underline{2}, \sigma), (\underline{1}, \rho), (\underline{2}, \rho), (\underline{2}, \rho^2), (\underline{1}, \sigma\rho^2) \right\}$$

$$\mathbf{B}_{264} = \left\{ (\underline{1}, 1), (\underline{2}, 1), (\underline{3}, 1), (\underline{2}, \sigma), (\underline{3}, \rho), (\underline{3}, \sigma\rho), (\underline{0}, \sigma\rho^2), (\underline{2}, \sigma\rho^2) \right\}$$

$$\mathbf{B}_{265} = \left\{ (\underline{1}, 1), (\underline{2}, 1), (\underline{3}, 1), (\underline{2}, \sigma), (\underline{0}, \sigma\rho), (\underline{1}, \sigma\rho), (\underline{1}, \rho^2), (\underline{3}, \sigma\rho^2) \right\}$$

$$\mathbf{B}_{266} = \left\{ (\underline{1}, 1), (\underline{2}, 1), (\underline{3}, 1), (\underline{3}, \sigma), (\underline{0}, \rho), (\underline{1}, \sigma\rho), (\underline{1}, \sigma\rho^2), (\underline{2}, \sigma\rho^2) \right\}$$

$$\mathbf{B}_{267} = \left\{ (\underline{1}, 1), (\underline{2}, 1), (\underline{3}, 1), (\underline{3}, \sigma), (\underline{1}, \rho), (\underline{3}, \sigma\rho), (\underline{3}, \rho^2), (\underline{3}, \sigma\rho^2) \right\}$$

$$\mathbf{B}_{268} = \left\{ (\underline{1}, 1), (\underline{2}, 1), (\underline{3}, 1), (\underline{3}, \sigma), (\underline{2}, \rho), (\underline{3}, \rho), (\underline{0}, \rho^2), (\underline{1}, \rho^2) \right\}$$

$$\mathbf{B}_{269} = \left\{ (\underline{1}, 1), (\underline{2}, 1), (\underline{3}, 1), (\underline{3}, \sigma), (\underline{0}, \sigma\rho), (\underline{2}, \sigma\rho), (\underline{2}, \rho^2), (\underline{0}, \sigma\rho^2) \right\}$$

$$\mathbf{B}_{270} = \left\{ (\underline{1}, 1), (\underline{2}, 1), (\underline{0}, \sigma), (\underline{1}, \sigma), (\underline{0}, \rho), (\underline{3}, \sigma\rho), (\underline{3}, \rho^2), (\underline{2}, \sigma\rho^2) \right\}$$

$$\mathbf{B}_{271} = \left\{ (\underline{1}, 1), (\underline{2}, 1), (\underline{0}, \sigma), (\underline{1}, \sigma), (\underline{1}, \rho), (\underline{1}, \sigma\rho), (\underline{1}, \sigma\rho^2), (\underline{3}, \sigma\rho^2) \right\}$$

$$\mathbf{B}_{272} = \left\{ (\underline{1}, 1), (\underline{2}, 1), (\underline{0}, \sigma), (\underline{1}, \sigma), (\underline{2}, \rho), (\underline{0}, \sigma\rho), (\underline{1}, \rho^2), (\underline{2}, \rho^2) \right\}$$

$$\mathbf{B}_{273} = \left\{ (\underline{1}, 1), (\underline{2}, 1), (\underline{0}, \sigma), (\underline{1}, \sigma), (\underline{3}, \rho), (\underline{2}, \sigma\rho), (\underline{0}, \rho^2), (\underline{0}, \sigma\rho^2) \right\}$$

$$\mathbf{B}_{274} = \left\{ (\underline{1}, 1), (\underline{2}, 1), (\underline{0}, \sigma), (\underline{2}, \sigma), (\underline{0}, \rho), (\underline{2}, \rho), (\underline{3}, \rho), (\underline{1}, \sigma\rho) \right\}$$

$$\mathbf{B}_{275} = \left\{ (\underline{1}, 1), (\underline{2}, 1), (\underline{0}, \sigma), (\underline{2}, \sigma), (\underline{1}, \rho), (\underline{0}, \sigma\rho), (\underline{2}, \sigma\rho), (\underline{3}, \sigma\rho) \right\}$$

$$\mathbf{B}_{276} = \left\{ (\underline{1}, 1), (\underline{2}, 1), (\underline{0}, \sigma), (\underline{2}, \sigma), (\underline{0}, \rho^2), (\underline{1}, \rho^2), (\underline{1}, \sigma\rho^2), (\underline{2}, \sigma\rho^2) \right\}$$

$$\mathbf{B}_{277} = \left\{ (\underline{1}, 1), (\underline{2}, 1), (\underline{0}, \sigma), (\underline{2}, \sigma), (\underline{2}, \rho^2), (\underline{3}, \rho^2), (\underline{0}, \sigma\rho^2), (\underline{3}, \sigma\rho^2) \right\}$$

$$\mathbf{B}_{278} = \left\{ (\underline{1}, 1), (\underline{2}, 1), (\underline{0}, \sigma), (\underline{3}, \sigma), (\underline{0}, \rho), (\underline{0}, \sigma\rho), (\underline{0}, \rho^2), (\underline{3}, \sigma\rho^2) \right\}$$

$$\mathbf{B}_{279} = \left\{ (\underline{1}, 1), (\underline{2}, 1), (\underline{0}, \sigma), (\underline{3}, \sigma), (\underline{1}, \rho), (\underline{3}, \rho), (\underline{2}, \rho^2), (\underline{2}, \sigma\rho^2) \right\}$$

$$\mathbf{B}_{280} = \left\{ (\underline{1}, 1), (\underline{2}, 1), (\underline{0}, \sigma), (\underline{3}, \sigma), (\underline{2}, \rho), (\underline{3}, \sigma\rho), (\underline{0}, \sigma\rho^2), (\underline{1}, \sigma\rho^2) \right\}$$

$$\mathbf{B}_{281} = \left\{ (\underline{1}, 1), (\underline{2}, 1), (\underline{0}, \sigma), (\underline{3}, \sigma), (\underline{1}, \sigma\rho), (\underline{2}, \sigma\rho), (\underline{1}, \rho^2), (\underline{3}, \rho^2) \right\}$$

$$\mathbf{B}_{282} = \left\{ (\underline{1}, 1), (\underline{2}, 1), (\underline{1}, \sigma), (\underline{2}, \sigma), (\underline{0}, \rho), (\underline{0}, \sigma\rho), (\underline{0}, \sigma\rho^2), (\underline{1}, \sigma\rho^2) \right\}$$

$$\mathbf{B}_{283} = \left\{ (\underline{1}, 1), (\underline{2}, 1), (\underline{1}, \sigma), (\underline{2}, \sigma), (\underline{1}, \rho), (\underline{3}, \rho), (\underline{1}, \rho^2), (\underline{3}, \rho^2) \right\}$$

$$\mathbf{B}_{284} = \left\{ (\underline{1}, 1), (\underline{2}, 1), (\underline{1}, \sigma), (\underline{2}, \sigma), (\underline{2}, \rho), (\underline{3}, \sigma\rho), (\underline{0}, \rho^2), (\underline{3}, \sigma\rho^2) \right\}$$

$$\mathbf{B}_{285} = \left\{ (\underline{1}, 1), (\underline{2}, 1), (\underline{1}, \sigma), (\underline{2}, \sigma), (\underline{1}, \sigma\rho), (\underline{2}, \sigma\rho), (\underline{2}, \rho^2), (\underline{2}, \sigma\rho^2) \right\}$$

$$\mathbf{B}_{286} = \left\{ (\underline{1}, 1), (\underline{2}, 1), (\underline{1}, \sigma), (\underline{3}, \sigma), (\underline{0}, \rho), (\underline{1}, \rho), (\underline{2}, \rho), (\underline{2}, \sigma\rho) \right\}$$

$$\mathbf{B}_{287} = \left\{ (\underline{1}, 1), (\underline{2}, 1), (\underline{1}, \sigma), (\underline{3}, \sigma), (\underline{3}, \rho), (\underline{0}, \sigma\rho), (\underline{1}, \sigma\rho), (\underline{3}, \sigma\rho) \right\}$$

$$\mathbf{B}_{288} = \left\{ (\underline{1}, 1), (\underline{2}, 1), (\underline{1}, \sigma), (\underline{3}, \sigma), (\underline{0}, \rho^2), (\underline{2}, \rho^2), (\underline{3}, \rho^2), (\underline{1}, \sigma\rho^2) \right\}$$

$$\mathbf{B}_{289} = \left\{ (\underline{1}, 1), (\underline{2}, 1), (\underline{1}, \sigma), (\underline{3}, \sigma), (\underline{1}, \rho^2), (\underline{0}, \sigma\rho^2), (\underline{2}, \sigma\rho^2), (\underline{3}, \sigma\rho^2) \right\}$$

$$\mathbf{B}_{290} = \left\{ (\underline{1}, 1), (\underline{2}, 1), (\underline{2}, \sigma), (\underline{3}, \sigma), (\underline{0}, \rho), (\underline{3}, \sigma\rho), (\underline{1}, \rho^2), (\underline{2}, \rho^2) \right\}$$

$$\mathbf{B}_{291} = \left\{ (\underline{1}, 1), (\underline{2}, 1), (\underline{2}, \sigma), (\underline{3}, \sigma), (\underline{1}, \rho), (\underline{1}, \sigma\rho), (\underline{0}, \rho^2), (\underline{0}, \sigma\rho^2) \right\}$$

$$\mathbf{B}_{292} = \left\{ (\underline{1}, 1), (\underline{2}, 1), (\underline{2}, \sigma), (\underline{3}, \sigma), (\underline{2}, \rho), (\underline{0}, \sigma\rho), (\underline{3}, \rho^2), (\underline{2}, \sigma\rho^2) \right\}$$

$$\mathbf{B}_{293} = \left\{ (\underline{1}, 1), (\underline{2}, 1), (\underline{2}, \sigma), (\underline{3}, \sigma), (\underline{3}, \rho), (\underline{2}, \sigma\rho), (\underline{1}, \sigma\rho^2), (\underline{3}, \sigma\rho^2) \right\}$$

$$\mathbf{B}_{294} = \left\{ (\underline{1}, 1), (\underline{2}, 1), (\underline{0}, \rho), (\underline{1}, \rho), (\underline{3}, \rho), (\underline{3}, \sigma\rho), (\underline{0}, \rho^2), (\underline{1}, \sigma\rho^2) \right\}$$

$$\mathbf{B}_{295} = \left\{ (\underline{1}, 1), (\underline{2}, 1), (\underline{0}, \rho), (\underline{1}, \rho), (\underline{0}, \sigma\rho), (\underline{1}, \sigma\rho), (\underline{2}, \rho^2), (\underline{3}, \rho^2) \right\}$$

$$\mathbf{B}_{296} = \left\{ (\underline{1}, 1), (\underline{2}, 1), (\underline{0}, \rho), (\underline{2}, \rho), (\underline{0}, \rho^2), (\underline{2}, \rho^2), (\underline{0}, \sigma\rho^2), (\underline{2}, \sigma\rho^2) \right\}$$

$$\mathbf{B}_{297} = \left\{ (\underline{1}, 1), (\underline{2}, 1), (\underline{0}, \rho), (\underline{2}, \rho), (\underline{1}, \rho^2), (\underline{3}, \rho^2), (\underline{1}, \sigma\rho^2), (\underline{3}, \sigma\rho^2) \right\}$$

$$\mathbf{B}_{298} = \left\{ (\underline{1}, 1), (\underline{2}, 1), (\underline{0}, \rho), (\underline{3}, \rho), (\underline{0}, \sigma\rho), (\underline{2}, \sigma\rho), (\underline{1}, \rho^2), (\underline{2}, \sigma\rho^2) \right\}$$

$$\mathbf{B}_{299} = \left\{ (\underline{1}, 1), (\underline{2}, 1), (\underline{0}, \rho), (\underline{1}, \sigma\rho), (\underline{2}, \sigma\rho), (\underline{3}, \sigma\rho), (\underline{0}, \sigma\rho^2), (\underline{3}, \sigma\rho^2) \right\}$$

$$\mathbf{B}_{300} = \left\{ (\underline{1}, 1), (\underline{2}, 1), (\underline{1}, \rho), (\underline{2}, \rho), (\underline{3}, \rho), (\underline{0}, \sigma\rho), (\underline{0}, \sigma\rho^2), (\underline{3}, \sigma\rho^2) \right\}$$

$$\mathbf{B}_{301} = \left\{ (\underline{1}, 1), (\underline{2}, 1), (\underline{1}, \rho), (\underline{2}, \rho), (\underline{1}, \sigma\rho), (\underline{3}, \sigma\rho), (\underline{1}, \rho^2), (\underline{2}, \sigma\rho^2) \right\}$$

$$\mathbf{B}_{302} = \left\{ (\underline{1}, 1), (\underline{2}, 1), (\underline{1}, \rho), (\underline{2}, \sigma\rho), (\underline{0}, \rho^2), (\underline{1}, \rho^2), (\underline{2}, \rho^2), (\underline{3}, \sigma\rho^2) \right\}$$

$$\mathbf{B}_{303} = \left\{ (\underline{1}, 1), (\underline{2}, 1), (\underline{1}, \rho), (\underline{2}, \sigma\rho), (\underline{3}, \rho^2), (\underline{0}, \sigma\rho^2), (\underline{1}, \sigma\rho^2), (\underline{2}, \sigma\rho^2) \right\}$$

$$\mathbf{B}_{304} = \left\{ (\underline{1}, 1), (\underline{2}, 1), (\underline{2}, \rho), (\underline{3}, \rho), (\underline{2}, \sigma\rho), (\underline{3}, \sigma\rho), (\underline{2}, \rho^2), (\underline{3}, \rho^2) \right\}$$

$$\mathbf{B}_{305} = \left\{ (\underline{1}, 1), (\underline{2}, 1), (\underline{2}, \rho), (\underline{0}, \sigma\rho), (\underline{1}, \sigma\rho), (\underline{2}, \sigma\rho), (\underline{0}, \rho^2), (\underline{1}, \sigma\rho^2) \right\}$$

$$\mathbf{B}_{306} = \left\{ (\underline{1}, 1), (\underline{2}, 1), (\underline{3}, \rho), (\underline{1}, \sigma\rho), (\underline{0}, \rho^2), (\underline{3}, \rho^2), (\underline{2}, \sigma\rho^2), (\underline{3}, \sigma\rho^2) \right\}$$

$$\mathbf{B}_{307} = \left\{ (\underline{1}, 1), (\underline{2}, 1), (\underline{3}, \rho), (\underline{1}, \sigma\rho), (\underline{1}, \rho^2), (\underline{2}, \rho^2), (\underline{0}, \sigma\rho^2), (\underline{1}, \sigma\rho^2) \right\}$$

$$\mathbf{B}_{308} = \left\{ (\underline{1}, 1), (\underline{2}, 1), (\underline{0}, \sigma\rho), (\underline{3}, \sigma\rho), (\underline{0}, \rho^2), (\underline{1}, \rho^2), (\underline{3}, \rho^2), (\underline{0}, \sigma\rho^2) \right\}$$

$$\mathbf{B}_{309} = \left\{ (\underline{1}, 1), (\underline{2}, 1), (\underline{0}, \sigma\rho), (\underline{3}, \sigma\rho), (\underline{2}, \rho^2), (\underline{1}, \sigma\rho^2), (\underline{2}, \sigma\rho^2), (\underline{3}, \sigma\rho^2) \right\}$$

$$\mathbf{B}_{310} = \left\{ (\underline{1}, 1), (\underline{3}, 1), (\underline{0}, \sigma), (\underline{1}, \sigma), (\underline{0}, \rho), (\underline{2}, \rho), (\underline{0}, \rho^2), (\underline{1}, \sigma\rho^2) \right\}$$

$$\mathbf{B}_{311} = \left\{ (\underline{1}, 1), (\underline{3}, 1), (\underline{0}, \sigma), (\underline{1}, \sigma), (\underline{1}, \rho), (\underline{2}, \sigma\rho), (\underline{2}, \rho^2), (\underline{3}, \rho^2) \right\}$$

$$\mathbf{B}_{312} = \left\{ (\underline{1}, 1), (\underline{3}, 1), (\underline{0}, \sigma), (\underline{1}, \sigma), (\underline{3}, \rho), (\underline{1}, \sigma\rho), (\underline{1}, \rho^2), (\underline{2}, \sigma\rho^2) \right\}$$

$$\mathbf{B}_{313} = \left\{ (\underline{1}, 1), (\underline{3}, 1), (\underline{0}, \sigma), (\underline{1}, \sigma), (\underline{0}, \sigma\rho), (\underline{3}, \sigma\rho), (\underline{0}, \sigma\rho^2), (\underline{3}, \sigma\rho^2) \right\}$$

$$\mathbf{B}_{314} = \left\{ (\underline{1}, 1), (\underline{3}, 1), (\underline{0}, \sigma), (\underline{2}, \sigma), (\underline{0}, \rho), (\underline{0}, \sigma\rho), (\underline{2}, \rho^2), (\underline{2}, \sigma\rho^2) \right\}$$

$$\mathbf{B}_{315} = \left\{ (\underline{1}, 1), (\underline{3}, 1), (\underline{0}, \sigma), (\underline{2}, \sigma), (\underline{1}, \rho), (\underline{3}, \rho), (\underline{0}, \rho^2), (\underline{3}, \sigma\rho^2) \right\}$$

$$\mathbf{B}_{316} = \left\{ (\underline{1}, 1), (\underline{3}, 1), (\underline{0}, \sigma), (\underline{2}, \sigma), (\underline{2}, \rho), (\underline{3}, \sigma\rho), (\underline{1}, \rho^2), (\underline{3}, \rho^2) \right\}$$

$$\mathbf{B}_{317} = \left\{ (\underline{1}, 1), (\underline{3}, 1), (\underline{0}, \sigma), (\underline{2}, \sigma), (\underline{1}, \sigma\rho), (\underline{2}, \sigma\rho), (\underline{0}, \sigma\rho^2), (\underline{1}, \sigma\rho^2) \right\}$$

$$\mathbf{B}_{318} = \left\{ (\underline{1}, 1), (\underline{3}, 1), (\underline{0}, \sigma), (\underline{3}, \sigma), (\underline{0}, \rho), (\underline{3}, \rho), (\underline{2}, \sigma\rho), (\underline{3}, \sigma\rho) \right\}$$

$$\mathbf{B}_{319} = \left\{ (\underline{1}, 1), (\underline{3}, 1), (\underline{0}, \sigma), (\underline{3}, \sigma), (\underline{1}, \rho), (\underline{2}, \rho), (\underline{0}, \sigma\rho), (\underline{1}, \sigma\rho) \right\}$$

$$\mathbf{B}_{320} = \left\{ (\underline{1}, 1), (\underline{3}, 1), (\underline{0}, \sigma), (\underline{3}, \sigma), (\underline{0}, \rho^2), (\underline{3}, \rho^2), (\underline{0}, \sigma\rho^2), (\underline{2}, \sigma\rho^2) \right\}$$

$$\mathbf{B}_{321} = \left\{ (\underline{1}, 1), (\underline{3}, 1), (\underline{0}, \sigma), (\underline{3}, \sigma), (\underline{1}, \rho^2), (\underline{2}, \rho^2), (\underline{1}, \sigma\rho^2), (\underline{3}, \sigma\rho^2) \right\}$$

$$\mathbf{B}_{322} = \left\{ (\underline{1}, 1), (\underline{3}, 1), (\underline{1}, \sigma), (\underline{2}, \sigma), (\underline{0}, \rho), (\underline{1}, \rho), (\underline{1}, \sigma\rho), (\underline{3}, \sigma\rho) \right\}$$

$$\mathbf{B}_{323} = \left\{ (\underline{1}, 1), (\underline{3}, 1), (\underline{1}, \sigma), (\underline{2}, \sigma), (\underline{2}, \rho), (\underline{3}, \rho), (\underline{0}, \sigma\rho), (\underline{2}, \sigma\rho) \right\}$$

$$\mathbf{B}_{324} = \left\{ (\underline{1}, 1), (\underline{3}, 1), (\underline{1}, \sigma), (\underline{2}, \sigma), (\underline{0}, \rho^2), (\underline{1}, \rho^2), (\underline{2}, \rho^2), (\underline{0}, \sigma\rho^2) \right\}$$

$$\mathbf{B}_{325} = \left\{ (\underline{1}, 1), (\underline{3}, 1), (\underline{1}, \sigma), (\underline{2}, \sigma), (\underline{3}, \rho^2), (\underline{1}, \sigma\rho^2), (\underline{2}, \sigma\rho^2), (\underline{3}, \sigma\rho^2) \right\}$$

$$\mathbf{B}_{326} = \left\{ (\underline{1}, 1), (\underline{3}, 1), (\underline{1}, \sigma), (\underline{3}, \sigma), (\underline{0}, \rho), (\underline{0}, \sigma\rho), (\underline{1}, \rho^2), (\underline{3}, \rho^2) \right\}$$

$$\mathbf{B}_{327} = \left\{ (\underline{1}, 1), (\underline{3}, 1), (\underline{1}, \sigma), (\underline{3}, \sigma), (\underline{1}, \rho), (\underline{3}, \rho), (\underline{0}, \sigma\rho^2), (\underline{1}, \sigma\rho^2) \right\}$$

$$\mathbf{B}_{328} = \left\{ (\underline{1}, 1), (\underline{3}, 1), (\underline{1}, \sigma), (\underline{3}, \sigma), (\underline{2}, \rho), (\underline{3}, \sigma\rho), (\underline{2}, \rho^2), (\underline{2}, \sigma\rho^2) \right\}$$

$$\mathbf{B}_{329} = \left\{ (\underline{1}, 1), (\underline{3}, 1), (\underline{1}, \sigma), (\underline{3}, \sigma), (\underline{1}, \sigma\rho), (\underline{2}, \sigma\rho), (\underline{0}, \rho^2), (\underline{3}, \sigma\rho^2) \right\}$$

$$\mathbf{B}_{330} = \left\{ (\underline{1}, 1), (\underline{3}, 1), (\underline{2}, \sigma), (\underline{3}, \sigma), (\underline{0}, \rho), (\underline{2}, \rho), (\underline{0}, \sigma\rho^2), (\underline{3}, \sigma\rho^2) \right\}$$

$$\mathbf{B}_{331} = \left\{ (\underline{1}, 1), (\underline{3}, 1), (\underline{2}, \sigma), (\underline{3}, \sigma), (\underline{1}, \rho), (\underline{2}, \sigma\rho), (\underline{1}, \rho^2), (\underline{2}, \sigma\rho^2) \right\}$$

$$\mathbf{B}_{332} = \left\{ (\underline{1}, 1), (\underline{3}, 1), (\underline{2}, \sigma), (\underline{3}, \sigma), (\underline{3}, \rho), (\underline{1}, \sigma\rho), (\underline{2}, \rho^2), (\underline{3}, \rho^2) \right\}$$

$$\mathbf{B}_{333} = \left\{ (\underline{1}, 1), (\underline{3}, 1), (\underline{2}, \sigma), (\underline{3}, \sigma), (\underline{0}, \sigma\rho), (\underline{3}, \sigma\rho), (\underline{0}, \rho^2), (\underline{1}, \sigma\rho^2) \right\}$$

$$\mathbf{B}_{334} = \left\{ (\underline{1}, 1), (\underline{3}, 1), (\underline{0}, \rho), (\underline{1}, \rho), (\underline{2}, \rho), (\underline{3}, \rho), (\underline{3}, \rho^2), (\underline{2}, \sigma\rho^2) \right\}$$

$$\mathbf{B}_{335} = \left\{ (\underline{1}, 1), (\underline{3}, 1), (\underline{0}, \rho), (\underline{1}, \rho), (\underline{0}, \sigma\rho), (\underline{2}, \sigma\rho), (\underline{1}, \sigma\rho^2), (\underline{3}, \sigma\rho^2) \right\}$$

$$\mathbf{B}_{336} = \left\{ (\underline{1}, 1), (\underline{3}, 1), (\underline{0}, \rho), (\underline{2}, \rho), (\underline{1}, \sigma\rho), (\underline{2}, \sigma\rho), (\underline{1}, \rho^2), (\underline{2}, \rho^2) \right\}$$

$$\mathbf{B}_{337} = \left\{ (\underline{1}, 1), (\underline{3}, 1), (\underline{0}, \rho), (\underline{3}, \rho), (\underline{0}, \sigma\rho), (\underline{1}, \sigma\rho), (\underline{0}, \rho^2), (\underline{0}, \sigma\rho^2) \right\}$$

$$\mathbf{B}_{338} = \left\{ (\underline{1}, 1), (\underline{3}, 1), (\underline{0}, \rho), (\underline{3}, \sigma\rho), (\underline{0}, \rho^2), (\underline{1}, \rho^2), (\underline{2}, \sigma\rho^2), (\underline{3}, \sigma\rho^2) \right\}$$

$$\mathbf{B}_{339} = \left\{ (\underline{1}, 1), (\underline{3}, 1), (\underline{0}, \rho), (\underline{3}, \sigma\rho), (\underline{2}, \rho^2), (\underline{3}, \rho^2), (\underline{0}, \sigma\rho^2), (\underline{1}, \sigma\rho^2) \right\}$$

$$\mathbf{B}_{340} = \left\{ (\underline{1}, 1), (\underline{3}, 1), (\underline{1}, \rho), (\underline{2}, \rho), (\underline{2}, \sigma\rho), (\underline{3}, \sigma\rho), (\underline{0}, \rho^2), (\underline{0}, \sigma\rho^2) \right\}$$

$$\mathbf{B}_{341} = \left\{ (\underline{1}, 1), (\underline{3}, 1), (\underline{1}, \rho), (\underline{3}, \rho), (\underline{0}, \sigma\rho), (\underline{3}, \sigma\rho), (\underline{1}, \rho^2), (\underline{2}, \rho^2) \right\}$$

$$\mathbf{B}_{342} = \left\{ (\underline{1}, 1), (\underline{3}, 1), (\underline{1}, \rho), (\underline{1}, \sigma\rho), (\underline{0}, \rho^2), (\underline{1}, \rho^2), (\underline{3}, \rho^2), (\underline{1}, \sigma\rho^2) \right\}$$

$$\mathbf{B}_{343} = \left\{ (\underline{1}, 1), (\underline{3}, 1), (\underline{1}, \rho), (\underline{1}, \sigma\rho), (\underline{2}, \rho^2), (\underline{0}, \sigma\rho^2), (\underline{2}, \sigma\rho^2), (\underline{3}, \sigma\rho^2) \right\}$$

$$\mathbf{B}_{344} = \left\{ (\underline{1}, 1), (\underline{3}, 1), (\underline{2}, \rho), (\underline{3}, \rho), (\underline{1}, \sigma\rho), (\underline{3}, \sigma\rho), (\underline{1}, \sigma\rho^2), (\underline{3}, \sigma\rho^2) \right\}$$

$$\mathbf{B}_{345} = \left\{ (\underline{1}, 1), (\underline{3}, 1), (\underline{2}, \rho), (\underline{0}, \sigma\rho), (\underline{0}, \rho^2), (\underline{2}, \rho^2), (\underline{3}, \rho^2), (\underline{3}, \sigma\rho^2) \right\}$$

$$\mathbf{B}_{346} = \left\{ (\underline{1}, 1), (\underline{3}, 1), (\underline{2}, \rho), (\underline{0}, \sigma\rho), (\underline{1}, \rho^2), (\underline{0}, \sigma\rho^2), (\underline{1}, \sigma\rho^2), (\underline{2}, \sigma\rho^2) \right\}$$

$$\mathbf{B}_{347} = \left\{ (\underline{1}, 1), (\underline{3}, 1), (\underline{3}, \rho), (\underline{2}, \sigma\rho), (\underline{0}, \rho^2), (\underline{2}, \rho^2), (\underline{1}, \sigma\rho^2), (\underline{2}, \sigma\rho^2) \right\}$$

$$\mathbf{B}_{348} = \left\{ (\underline{1}, 1), (\underline{3}, 1), (\underline{3}, \rho), (\underline{2}, \sigma\rho), (\underline{1}, \rho^2), (\underline{3}, \rho^2), (\underline{0}, \sigma\rho^2), (\underline{3}, \sigma\rho^2) \right\}$$

$$\mathbf{B}_{349} = \left\{ (\underline{1}, 1), (\underline{3}, 1), (\underline{0}, \sigma\rho), (\underline{1}, \sigma\rho), (\underline{2}, \sigma\rho), (\underline{3}, \sigma\rho), (\underline{3}, \rho^2), (\underline{2}, \sigma\rho^2) \right\}$$

$$\mathbf{B}_{350} = \left\{ (\underline{1}, 1), (\underline{0}, \sigma), (\underline{1}, \sigma), (\underline{2}, \sigma), (\underline{0}, \rho), (\underline{2}, \sigma\rho), (\underline{1}, \rho^2), (\underline{3}, \sigma\rho^2) \right\}$$

$$\mathbf{B}_{351} = \left\{ (\underline{1}, 1), (\underline{0}, \sigma), (\underline{1}, \sigma), (\underline{2}, \sigma), (\underline{1}, \rho), (\underline{2}, \rho), (\underline{0}, \sigma\rho^2), (\underline{2}, \sigma\rho^2) \right\}$$

$$\mathbf{B}_{352} = \left\{ (\underline{1}, 1), (\underline{0}, \sigma), (\underline{1}, \sigma), (\underline{2}, \sigma), (\underline{3}, \rho), (\underline{3}, \sigma\rho), (\underline{2}, \rho^2), (\underline{1}, \sigma\rho^2) \right\}$$

$$\mathbf{B}_{353} = \left\{ (\underline{1}, 1), (\underline{0}, \sigma), (\underline{1}, \sigma), (\underline{2}, \sigma), (\underline{0}, \sigma\rho), (\underline{1}, \sigma\rho), (\underline{0}, \rho^2), (\underline{3}, \rho^2) \right\}$$

$$\mathbf{B}_{354} = \left\{ (\underline{1}, 1), (\underline{0}, \sigma), (\underline{1}, \sigma), (\underline{3}, \sigma), (\underline{0}, \rho), (\underline{1}, \sigma\rho), (\underline{2}, \rho^2), (\underline{0}, \sigma\rho^2) \right\}$$

$$\mathbf{B}_{355} = \left\{ (\underline{1}, 1), (\underline{0}, \sigma), (\underline{1}, \sigma), (\underline{3}, \sigma), (\underline{1}, \rho), (\underline{3}, \sigma\rho), (\underline{0}, \rho^2), (\underline{1}, \rho^2) \right\}$$

$$\mathbf{B}_{356} = \left\{ (\underline{1}, 1), (\underline{0}, \sigma), (\underline{1}, \sigma), (\underline{3}, \sigma), (\underline{2}, \rho), (\underline{3}, \rho), (\underline{3}, \rho^2), (\underline{3}, \sigma\rho^2) \right\}$$

$$\mathbf{B}_{357} = \left\{ (\underline{1}, 1), (\underline{0}, \sigma), (\underline{1}, \sigma), (\underline{3}, \sigma), (\underline{0}, \sigma\rho), (\underline{2}, \sigma\rho), (\underline{1}, \sigma\rho^2), (\underline{2}, \sigma\rho^2) \right\}$$

$$\mathbf{B}_{358} = \left\{ (\underline{1}, 1), (\underline{0}, \sigma), (\underline{2}, \sigma), (\underline{3}, \sigma), (\underline{0}, \rho), (\underline{1}, \rho), (\underline{3}, \rho^2), (\underline{1}, \sigma\rho^2) \right\}$$

$$\mathbf{B}_{359} = \left\{ (\underline{1}, 1), (\underline{0}, \sigma), (\underline{2}, \sigma), (\underline{3}, \sigma), (\underline{2}, \rho), (\underline{2}, \sigma\rho), (\underline{0}, \rho^2), (\underline{2}, \rho^2) \right\}$$

$$\mathbf{B}_{360} = \left\{ (\underline{1}, 1), (\underline{0}, \sigma), (\underline{2}, \sigma), (\underline{3}, \sigma), (\underline{3}, \rho), (\underline{0}, \sigma\rho), (\underline{1}, \rho^2), (\underline{0}, \sigma\rho^2) \right\}$$

$$\mathbf{B}_{361} = \left\{ (\underline{1}, 1), (\underline{0}, \sigma), (\underline{2}, \sigma), (\underline{3}, \sigma), (\underline{1}, \sigma\rho), (\underline{3}, \sigma\rho), (\underline{2}, \sigma\rho^2), (\underline{3}, \sigma\rho^2) \right\}$$

$$\mathbf{B}_{362} = \left\{ (\underline{1}, 1), (\underline{0}, \sigma), (\underline{0}, \rho), (\underline{1}, \rho), (\underline{2}, \rho), (\underline{3}, \sigma\rho), (\underline{2}, \rho^2), (\underline{3}, \sigma\rho^2) \right\}$$

$$\mathbf{B}_{363} = \left\{ (\underline{1}, 1), (\underline{0}, \sigma), (\underline{0}, \rho), (\underline{1}, \rho), (\underline{1}, \sigma\rho), (\underline{2}, \sigma\rho), (\underline{0}, \rho^2), (\underline{2}, \sigma\rho^2) \right\}$$

$$\mathbf{B}_{364} = \left\{ (\underline{1}, 1), (\underline{0}, \sigma), (\underline{0}, \rho), (\underline{2}, \rho), (\underline{0}, \sigma\rho), (\underline{2}, \sigma\rho), (\underline{3}, \rho^2), (\underline{0}, \sigma\rho^2) \right\}$$

$$\mathbf{B}_{365} = \left\{ (\underline{1}, 1), (\underline{0}, \sigma), (\underline{0}, \rho), (\underline{3}, \rho), (\underline{0}, \rho^2), (\underline{1}, \rho^2), (\underline{2}, \rho^2), (\underline{3}, \rho^2) \right\}$$

$$\mathbf{B}_{366} = \left\{ (\underline{1}, 1), (\underline{0}, \sigma), (\underline{0}, \rho), (\underline{3}, \rho), (\underline{0}, \sigma\rho^2), (\underline{1}, \sigma\rho^2), (\underline{2}, \sigma\rho^2), (\underline{3}, \sigma\rho^2) \right\}$$

$$\mathbf{B}_{367} = \left\{ (\underline{1}, 1), (\underline{0}, \sigma), (\underline{0}, \rho), (\underline{0}, \sigma\rho), (\underline{1}, \sigma\rho), (\underline{3}, \sigma\rho), (\underline{1}, \rho^2), (\underline{1}, \sigma\rho^2) \right\}$$

$$\mathbf{B}_{368} = \left\{ (\underline{1}, 1), (\underline{0}, \sigma), (\underline{1}, \rho), (\underline{2}, \rho), (\underline{3}, \rho), (\underline{2}, \sigma\rho), (\underline{1}, \rho^2), (\underline{1}, \sigma\rho^2) \right\}$$

$$\mathbf{B}_{369} = \left\{ (\underline{1}, 1), (\underline{0}, \sigma), (\underline{1}, \rho), (\underline{3}, \rho), (\underline{1}, \sigma\rho), (\underline{3}, \sigma\rho), (\underline{3}, \rho^2), (\underline{0}, \sigma\rho^2) \right\}$$

$$\mathbf{B}_{370} = \left\{ (\underline{1}, 1), (\underline{0}, \sigma), (\underline{1}, \rho), (\underline{0}, \sigma\rho), (\underline{0}, \rho^2), (\underline{2}, \rho^2), (\underline{0}, \sigma\rho^2), (\underline{1}, \sigma\rho^2) \right\}$$

$$\mathbf{B}_{371} = \left\{ (\underline{1}, 1), (\underline{0}, \sigma), (\underline{1}, \rho), (\underline{0}, \sigma\rho), (\underline{1}, \rho^2), (\underline{3}, \rho^2), (\underline{2}, \sigma\rho^2), (\underline{3}, \sigma\rho^2) \right\}$$

$$\mathbf{B}_{372} = \left\{ (\underline{1}, 1), (\underline{0}, \sigma), (\underline{2}, \rho), (\underline{3}, \rho), (\underline{0}, \sigma\rho), (\underline{3}, \sigma\rho), (\underline{0}, \rho^2), (\underline{2}, \sigma\rho^2) \right\}$$

$$\mathbf{B}_{373} = \left\{ (\underline{1}, 1), (\underline{0}, \sigma), (\underline{2}, \rho), (\underline{1}, \sigma\rho), (\underline{0}, \rho^2), (\underline{1}, \rho^2), (\underline{0}, \sigma\rho^2), (\underline{3}, \sigma\rho^2) \right\}$$

$$\mathbf{B}_{374} = \left\{ (\underline{1}, 1), (\underline{0}, \sigma), (\underline{2}, \rho), (\underline{1}, \sigma\rho), (\underline{2}, \rho^2), (\underline{3}, \rho^2), (\underline{1}, \sigma\rho^2), (\underline{2}, \sigma\rho^2) \right\}$$

$$\mathbf{B}_{375} = \left\{ (\underline{1}, 1), (\underline{0}, \sigma), (\underline{3}, \rho), (\underline{0}, \sigma\rho), (\underline{1}, \sigma\rho), (\underline{2}, \sigma\rho), (\underline{2}, \rho^2), (\underline{3}, \sigma\rho^2) \right\}$$

$$\mathbf{B}_{376} = \left\{ (\underline{1}, 1), (\underline{0}, \sigma), (\underline{2}, \sigma\rho), (\underline{3}, \sigma\rho), (\underline{0}, \rho^2), (\underline{3}, \rho^2), (\underline{1}, \sigma\rho^2), (\underline{3}, \sigma\rho^2) \right\}$$

$$\mathbf{B}_{377} = \left\{ (\underline{1}, 1), (\underline{0}, \sigma), (\underline{2}, \sigma\rho), (\underline{3}, \sigma\rho), (\underline{1}, \rho^2), (\underline{2}, \rho^2), (\underline{0}, \sigma\rho^2), (\underline{2}, \sigma\rho^2) \right\}$$

$$\mathbf{B}_{378} = \left\{ (\underline{1}, 1), (\underline{1}, \sigma), (\underline{2}, \sigma), (\underline{3}, \sigma), (\underline{0}, \rho), (\underline{3}, \rho), (\underline{0}, \rho^2), (\underline{2}, \sigma\rho^2) \right\}$$

$$\mathbf{B}_{379} = \left\{ (\underline{1}, 1), (\underline{1}, \sigma), (\underline{2}, \sigma), (\underline{3}, \sigma), (\underline{1}, \rho), (\underline{0}, \sigma\rho), (\underline{2}, \rho^2), (\underline{3}, \sigma\rho^2) \right\}$$

$$\mathbf{B}_{380} = \{ (\underline{1}, 1), (\underline{1}, \sigma), (\underline{2}, \sigma), (\underline{3}, \sigma), (\underline{2}, \rho), (\underline{1}, \sigma\rho), (\underline{1}, \rho^2), (\underline{1}, \sigma\rho^2) \}$$

$$\mathbf{B}_{381} = \{ (\underline{1}, 1), (\underline{1}, \sigma), (\underline{2}, \sigma), (\underline{3}, \sigma), (\underline{2}, \sigma\rho), (\underline{3}, \sigma\rho), (\underline{3}, \rho^2), (\underline{0}, \sigma\rho^2) \}$$

$$\mathbf{B}_{382} = \{ (\underline{1}, 1), (\underline{1}, \sigma), (\underline{0}, \rho), (\underline{1}, \rho), (\underline{0}, \rho^2), (\underline{3}, \rho^2), (\underline{0}, \sigma\rho^2), (\underline{3}, \sigma\rho^2) \}$$

$$\mathbf{B}_{383} = \{ (\underline{1}, 1), (\underline{1}, \sigma), (\underline{0}, \rho), (\underline{1}, \rho), (\underline{1}, \rho^2), (\underline{2}, \rho^2), (\underline{1}, \sigma\rho^2), (\underline{2}, \sigma\rho^2) \}$$

$$\mathbf{B}_{384} = \{ (\underline{1}, 1), (\underline{1}, \sigma), (\underline{0}, \rho), (\underline{2}, \rho), (\underline{3}, \rho), (\underline{3}, \sigma\rho), (\underline{1}, \rho^2), (\underline{0}, \sigma\rho^2) \}$$

$$\mathbf{B}_{385} = \{ (\underline{1}, 1), (\underline{1}, \sigma), (\underline{0}, \rho), (\underline{2}, \rho), (\underline{0}, \sigma\rho), (\underline{1}, \sigma\rho), (\underline{2}, \sigma\rho^2), (\underline{3}, \sigma\rho^2) \}$$

$$\mathbf{B}_{386} = \{ (\underline{1}, 1), (\underline{1}, \sigma), (\underline{0}, \rho), (\underline{3}, \rho), (\underline{1}, \sigma\rho), (\underline{2}, \sigma\rho), (\underline{3}, \rho^2), (\underline{1}, \sigma\rho^2) \}$$

$$\mathbf{B}_{387} = \{ (\underline{1}, 1), (\underline{1}, \sigma), (\underline{0}, \rho), (\underline{0}, \sigma\rho), (\underline{2}, \sigma\rho), (\underline{3}, \sigma\rho), (\underline{0}, \rho^2), (\underline{2}, \rho^2) \}$$

$$\mathbf{B}_{388} = \{ (\underline{1}, 1), (\underline{1}, \sigma), (\underline{1}, \rho), (\underline{2}, \rho), (\underline{3}, \rho), (\underline{1}, \sigma\rho), (\underline{0}, \rho^2), (\underline{2}, \rho^2) \}$$

$$\mathbf{B}_{389} = \{ (\underline{1}, 1), (\underline{1}, \sigma), (\underline{1}, \rho), (\underline{2}, \rho), (\underline{0}, \sigma\rho), (\underline{3}, \sigma\rho), (\underline{3}, \rho^2), (\underline{1}, \sigma\rho^2) \}$$

$$\mathbf{B}_{390} = \{ (\underline{1}, 1), (\underline{1}, \sigma), (\underline{1}, \rho), (\underline{3}, \rho), (\underline{2}, \sigma\rho), (\underline{3}, \sigma\rho), (\underline{2}, \sigma\rho^2), (\underline{3}, \sigma\rho^2) \}$$

$$\mathbf{B}_{391} = \{ (\underline{1}, 1), (\underline{1}, \sigma), (\underline{1}, \rho), (\underline{0}, \sigma\rho), (\underline{1}, \sigma\rho), (\underline{2}, \sigma\rho), (\underline{1}, \rho^2), (\underline{0}, \sigma\rho^2) \}$$

$$\mathbf{B}_{392} = \{ (\underline{1}, 1), (\underline{1}, \sigma), (\underline{2}, \rho), (\underline{2}, \sigma\rho), (\underline{0}, \rho^2), (\underline{1}, \rho^2), (\underline{3}, \rho^2), (\underline{2}, \sigma\rho^2) \}$$

$$\mathbf{B}_{393} = \{ (\underline{1}, 1), (\underline{1}, \sigma), (\underline{2}, \rho), (\underline{2}, \sigma\rho), (\underline{2}, \rho^2), (\underline{0}, \sigma\rho^2), (\underline{1}, \sigma\rho^2), (\underline{3}, \sigma\rho^2) \}$$

$$\mathbf{B}_{394} = \{ (\underline{1}, 1), (\underline{1}, \sigma), (\underline{3}, \rho), (\underline{0}, \sigma\rho), (\underline{0}, \rho^2), (\underline{1}, \rho^2), (\underline{1}, \sigma\rho^2), (\underline{3}, \sigma\rho^2) \}$$

$$\mathbf{B}_{395} = \{ (\underline{1}, 1), (\underline{1}, \sigma), (\underline{3}, \rho), (\underline{0}, \sigma\rho), (\underline{2}, \rho^2), (\underline{3}, \rho^2), (\underline{0}, \sigma\rho^2), (\underline{2}, \sigma\rho^2) \}$$

$$\mathbf{B}_{396} = \{ (\underline{1}, 1), (\underline{1}, \sigma), (\underline{1}, \sigma\rho), (\underline{3}, \sigma\rho), (\underline{0}, \rho^2), (\underline{0}, \sigma\rho^2), (\underline{1}, \sigma\rho^2), (\underline{2}, \sigma\rho^2) \}$$

$$\mathbf{B}_{397} = \{ (\underline{1}, 1), (\underline{1}, \sigma), (\underline{1}, \sigma\rho), (\underline{3}, \sigma\rho), (\underline{1}, \rho^2), (\underline{2}, \rho^2), (\underline{3}, \rho^2), (\underline{3}, \sigma\rho^2) \}$$

$$\mathbf{B}_{398} = \{ (\underline{1}, 1), (\underline{2}, \sigma), (\underline{0}, \rho), (\underline{1}, \rho), (\underline{2}, \rho), (\underline{0}, \sigma\rho), (\underline{0}, \rho^2), (\underline{1}, \rho^2) \}$$

$$\mathbf{B}_{399} = \{ (\underline{1}, 1), (\underline{2}, \sigma), (\underline{0}, \rho), (\underline{1}, \rho), (\underline{3}, \rho), (\underline{2}, \sigma\rho), (\underline{2}, \rho^2), (\underline{0}, \sigma\rho^2) \}$$

$$\mathbf{B}_{400} = \{ (\underline{1}, 1), (\underline{2}, \sigma), (\underline{0}, \rho), (\underline{2}, \rho), (\underline{2}, \sigma\rho), (\underline{3}, \sigma\rho), (\underline{1}, \sigma\rho^2), (\underline{2}, \sigma\rho^2) \}$$

$$\mathbf{B}_{401} = \{ (\underline{1}, 1), (\underline{2}, \sigma), (\underline{0}, \rho), (\underline{3}, \rho), (\underline{0}, \sigma\rho), (\underline{3}, \sigma\rho), (\underline{3}, \rho^2), (\underline{3}, \sigma\rho^2) \}$$

$\mathbf{B}_{402} = \{ (\underline{1}, 1), (\underline{2}, \sigma), (\underline{0}, \rho), (\underline{1}, \sigma\rho), (\underline{0}, \rho^2), (\underline{2}, \rho^2), (\underline{1}, \sigma\rho^2), (\underline{3}, \sigma\rho^2) \}$

$\mathbf{B}_{403} = \{ (\underline{1}, 1), (\underline{2}, \sigma), (\underline{0}, \rho), (\underline{1}, \sigma\rho), (\underline{1}, \rho^2), (\underline{3}, \rho^2), (\underline{0}, \sigma\rho^2), (\underline{2}, \sigma\rho^2) \}$

$\mathbf{B}_{404} = \{ (\underline{1}, 1), (\underline{2}, \sigma), (\underline{1}, \rho), (\underline{2}, \rho), (\underline{1}, \sigma\rho), (\underline{2}, \sigma\rho), (\underline{3}, \rho^2), (\underline{3}, \sigma\rho^2) \}$

$\mathbf{B}_{405} = \{ (\underline{1}, 1), (\underline{2}, \sigma), (\underline{1}, \rho), (\underline{3}, \rho), (\underline{0}, \sigma\rho), (\underline{1}, \sigma\rho), (\underline{1}, \sigma\rho^2), (\underline{2}, \sigma\rho^2) \}$

$\mathbf{B}_{406} = \{ (\underline{1}, 1), (\underline{2}, \sigma), (\underline{1}, \rho), (\underline{3}, \sigma\rho), (\underline{0}, \rho^2), (\underline{2}, \rho^2), (\underline{3}, \rho^2), (\underline{2}, \sigma\rho^2) \}$

$\mathbf{B}_{407} = \{ (\underline{1}, 1), (\underline{2}, \sigma), (\underline{1}, \rho), (\underline{3}, \sigma\rho), (\underline{1}, \rho^2), (\underline{0}, \sigma\rho^2), (\underline{1}, \sigma\rho^2), (\underline{3}, \sigma\rho^2) \}$

$\mathbf{B}_{408} = \{ (\underline{1}, 1), (\underline{2}, \sigma), (\underline{2}, \rho), (\underline{3}, \rho), (\underline{0}, \rho^2), (\underline{3}, \rho^2), (\underline{0}, \sigma\rho^2), (\underline{1}, \sigma\rho^2) \}$

$\mathbf{B}_{409} = \{ (\underline{1}, 1), (\underline{2}, \sigma), (\underline{2}, \rho), (\underline{3}, \rho), (\underline{1}, \rho^2), (\underline{2}, \rho^2), (\underline{2}, \sigma\rho^2), (\underline{3}, \sigma\rho^2) \}$

$\mathbf{B}_{410} = \{ (\underline{1}, 1), (\underline{2}, \sigma), (\underline{2}, \rho), (\underline{0}, \sigma\rho), (\underline{1}, \sigma\rho), (\underline{3}, \sigma\rho), (\underline{2}, \rho^2), (\underline{0}, \sigma\rho^2) \}$

$\mathbf{B}_{411} = \{ (\underline{1}, 1), (\underline{2}, \sigma), (\underline{3}, \rho), (\underline{1}, \sigma\rho), (\underline{2}, \sigma\rho), (\underline{3}, \sigma\rho), (\underline{0}, \rho^2), (\underline{1}, \rho^2) \}$

$\mathbf{B}_{412} = \{ (\underline{1}, 1), (\underline{2}, \sigma), (\underline{0}, \sigma\rho), (\underline{2}, \sigma\rho), (\underline{0}, \rho^2), (\underline{0}, \sigma\rho^2), (\underline{2}, \sigma\rho^2), (\underline{3}, \sigma\rho^2) \}$

$\mathbf{B}_{413} = \{ (\underline{1}, 1), (\underline{2}, \sigma), (\underline{0}, \sigma\rho), (\underline{2}, \sigma\rho), (\underline{1}, \rho^2), (\underline{2}, \rho^2), (\underline{3}, \rho^2), (\underline{1}, \sigma\rho^2) \}$

$\mathbf{B}_{414} = \{ (\underline{1}, 1), (\underline{3}, \sigma), (\underline{0}, \rho), (\underline{1}, \rho), (\underline{3}, \rho), (\underline{1}, \sigma\rho), (\underline{1}, \rho^2), (\underline{3}, \sigma\rho^2) \}$

$\mathbf{B}_{415} = \{ (\underline{1}, 1), (\underline{3}, \sigma), (\underline{0}, \rho), (\underline{1}, \rho), (\underline{0}, \sigma\rho), (\underline{3}, \sigma\rho), (\underline{0}, \sigma\rho^2), (\underline{2}, \sigma\rho^2) \}$

$\mathbf{B}_{416} = \{ (\underline{1}, 1), (\underline{3}, \sigma), (\underline{0}, \rho), (\underline{2}, \rho), (\underline{3}, \rho), (\underline{0}, \sigma\rho), (\underline{2}, \rho^2), (\underline{1}, \sigma\rho^2) \}$

$\mathbf{B}_{417} = \{ (\underline{1}, 1), (\underline{3}, \sigma), (\underline{0}, \rho), (\underline{2}, \rho), (\underline{1}, \sigma\rho), (\underline{3}, \sigma\rho), (\underline{0}, \rho^2), (\underline{3}, \rho^2) \}$

$\mathbf{B}_{418} = \{ (\underline{1}, 1), (\underline{3}, \sigma), (\underline{0}, \rho), (\underline{2}, \sigma\rho), (\underline{0}, \rho^2), (\underline{1}, \rho^2), (\underline{0}, \sigma\rho^2), (\underline{1}, \sigma\rho^2) \}$

$\mathbf{B}_{419} = \{ (\underline{1}, 1), (\underline{3}, \sigma), (\underline{0}, \rho), (\underline{2}, \sigma\rho), (\underline{2}, \rho^2), (\underline{3}, \rho^2), (\underline{2}, \sigma\rho^2), (\underline{3}, \sigma\rho^2) \}$

$\mathbf{B}_{420} = \{ (\underline{1}, 1), (\underline{3}, \sigma), (\underline{1}, \rho), (\underline{2}, \rho), (\underline{0}, \rho^2), (\underline{1}, \sigma\rho^2), (\underline{2}, \sigma\rho^2), (\underline{3}, \sigma\rho^2) \}$

$\mathbf{B}_{421} = \{ (\underline{1}, 1), (\underline{3}, \sigma), (\underline{1}, \rho), (\underline{2}, \rho), (\underline{1}, \rho^2), (\underline{2}, \rho^2), (\underline{3}, \rho^2), (\underline{0}, \sigma\rho^2) \}$

$\mathbf{B}_{422} = \{ (\underline{1}, 1), (\underline{3}, \sigma), (\underline{1}, \rho), (\underline{3}, \rho), (\underline{0}, \sigma\rho), (\underline{2}, \sigma\rho), (\underline{0}, \rho^2), (\underline{3}, \rho^2) \}$

$\mathbf{B}_{423} = \{ (\underline{1}, 1), (\underline{3}, \sigma), (\underline{1}, \rho), (\underline{1}, \sigma\rho), (\underline{2}, \sigma\rho), (\underline{3}, \sigma\rho), (\underline{2}, \rho^2), (\underline{1}, \sigma\rho^2) \}$

$$\mathbf{B}_{424} = \left\{ (\underline{1}, 1), (\underline{3}, \sigma), (\underline{2}, \rho), (\underline{3}, \rho), (\underline{1}, \sigma\rho), (\underline{2}, \sigma\rho), (\underline{0}, \sigma\rho^2), (\underline{2}, \sigma\rho^2) \right\}$$

$$\mathbf{B}_{425} = \left\{ (\underline{1}, 1), (\underline{3}, \sigma), (\underline{2}, \rho), (\underline{0}, \sigma\rho), (\underline{2}, \sigma\rho), (\underline{3}, \sigma\rho), (\underline{1}, \rho^2), (\underline{3}, \sigma\rho^2) \right\}$$

$$\mathbf{B}_{426} = \left\{ (\underline{1}, 1), (\underline{3}, \sigma), (\underline{3}, \rho), (\underline{3}, \sigma\rho), (\underline{0}, \rho^2), (\underline{2}, \rho^2), (\underline{0}, \sigma\rho^2), (\underline{3}, \sigma\rho^2) \right\}$$

$$\mathbf{B}_{427} = \left\{ (\underline{1}, 1), (\underline{3}, \sigma), (\underline{3}, \rho), (\underline{3}, \sigma\rho), (\underline{1}, \rho^2), (\underline{3}, \rho^2), (\underline{1}, \sigma\rho^2), (\underline{2}, \sigma\rho^2) \right\}$$

$$\mathbf{B}_{428} = \left\{ (\underline{1}, 1), (\underline{3}, \sigma), (\underline{0}, \sigma\rho), (\underline{1}, \sigma\rho), (\underline{0}, \rho^2), (\underline{1}, \rho^2), (\underline{2}, \rho^2), (\underline{2}, \sigma\rho^2) \right\}$$

$$\mathbf{B}_{429} = \left\{ (\underline{1}, 1), (\underline{3}, \sigma), (\underline{0}, \sigma\rho), (\underline{1}, \sigma\rho), (\underline{3}, \rho^2), (\underline{0}, \sigma\rho^2), (\underline{1}, \sigma\rho^2), (\underline{3}, \sigma\rho^2) \right\}$$

$$\mathbf{B}_{430} = \left\{ (\underline{2}, 1), (\underline{3}, 1), (\underline{0}, \sigma), (\underline{1}, \sigma), (\underline{0}, \rho), (\underline{0}, \sigma\rho), (\underline{1}, \sigma\rho), (\underline{2}, \sigma\rho) \right\}$$

$$\mathbf{B}_{431} = \left\{ (\underline{2}, 1), (\underline{3}, 1), (\underline{0}, \sigma), (\underline{1}, \sigma), (\underline{1}, \rho), (\underline{2}, \rho), (\underline{3}, \rho), (\underline{3}, \sigma\rho) \right\}$$

$$\mathbf{B}_{432} = \left\{ (\underline{2}, 1), (\underline{3}, 1), (\underline{0}, \sigma), (\underline{1}, \sigma), (\underline{0}, \rho^2), (\underline{1}, \rho^2), (\underline{3}, \rho^2), (\underline{3}, \sigma\rho^2) \right\}$$

$$\mathbf{B}_{433} = \left\{ (\underline{2}, 1), (\underline{3}, 1), (\underline{0}, \sigma), (\underline{1}, \sigma), (\underline{2}, \rho^2), (\underline{0}, \sigma\rho^2), (\underline{1}, \sigma\rho^2), (\underline{2}, \sigma\rho^2) \right\}$$

$$\mathbf{B}_{434} = \left\{ (\underline{2}, 1), (\underline{3}, 1), (\underline{0}, \sigma), (\underline{2}, \sigma), (\underline{0}, \rho), (\underline{3}, \sigma\rho), (\underline{1}, \sigma\rho^2), (\underline{3}, \sigma\rho^2) \right\}$$

$$\mathbf{B}_{435} = \left\{ (\underline{2}, 1), (\underline{3}, 1), (\underline{0}, \sigma), (\underline{2}, \sigma), (\underline{1}, \rho), (\underline{1}, \sigma\rho), (\underline{3}, \rho^2), (\underline{2}, \sigma\rho^2) \right\}$$

$$\mathbf{B}_{436} = \left\{ (\underline{2}, 1), (\underline{3}, 1), (\underline{0}, \sigma), (\underline{2}, \sigma), (\underline{2}, \rho), (\underline{0}, \sigma\rho), (\underline{0}, \rho^2), (\underline{0}, \sigma\rho^2) \right\}$$

$$\mathbf{B}_{437} = \left\{ (\underline{2}, 1), (\underline{3}, 1), (\underline{0}, \sigma), (\underline{2}, \sigma), (\underline{3}, \rho), (\underline{2}, \sigma\rho), (\underline{1}, \rho^2), (\underline{2}, \rho^2) \right\}$$

$$\mathbf{B}_{438} = \left\{ (\underline{2}, 1), (\underline{3}, 1), (\underline{0}, \sigma), (\underline{3}, \sigma), (\underline{0}, \rho), (\underline{2}, \rho), (\underline{2}, \rho^2), (\underline{3}, \rho^2) \right\}$$

$$\mathbf{B}_{439} = \left\{ (\underline{2}, 1), (\underline{3}, 1), (\underline{0}, \sigma), (\underline{3}, \sigma), (\underline{1}, \rho), (\underline{2}, \sigma\rho), (\underline{0}, \rho^2), (\underline{1}, \sigma\rho^2) \right\}$$

$$\mathbf{B}_{440} = \left\{ (\underline{2}, 1), (\underline{3}, 1), (\underline{0}, \sigma), (\underline{3}, \sigma), (\underline{3}, \rho), (\underline{1}, \sigma\rho), (\underline{0}, \sigma\rho^2), (\underline{3}, \sigma\rho^2) \right\}$$

$$\mathbf{B}_{441} = \left\{ (\underline{2}, 1), (\underline{3}, 1), (\underline{0}, \sigma), (\underline{3}, \sigma), (\underline{0}, \sigma\rho), (\underline{3}, \sigma\rho), (\underline{1}, \rho^2), (\underline{2}, \sigma\rho^2) \right\}$$

$$\mathbf{B}_{442} = \left\{ (\underline{2}, 1), (\underline{3}, 1), (\underline{1}, \sigma), (\underline{2}, \sigma), (\underline{0}, \rho), (\underline{2}, \rho), (\underline{1}, \rho^2), (\underline{2}, \sigma\rho^2) \right\}$$

$$\mathbf{B}_{443} = \left\{ (\underline{2}, 1), (\underline{3}, 1), (\underline{1}, \sigma), (\underline{2}, \sigma), (\underline{1}, \rho), (\underline{2}, \sigma\rho), (\underline{0}, \sigma\rho^2), (\underline{3}, \sigma\rho^2) \right\}$$

$$\mathbf{B}_{444} = \left\{ (\underline{2}, 1), (\underline{3}, 1), (\underline{1}, \sigma), (\underline{2}, \sigma), (\underline{3}, \rho), (\underline{1}, \sigma\rho), (\underline{0}, \rho^2), (\underline{1}, \sigma\rho^2) \right\}$$

$$\mathbf{B}_{445} = \left\{ (\underline{2}, 1), (\underline{3}, 1), (\underline{1}, \sigma), (\underline{2}, \sigma), (\underline{0}, \sigma\rho), (\underline{3}, \sigma\rho), (\underline{2}, \rho^2), (\underline{3}, \rho^2) \right\}$$

$$\mathbf{B}_{446} = \left\{ (\underline{2}, 1), (\underline{3}, 1), (\underline{1}, \sigma), (\underline{3}, \sigma), (\underline{0}, \rho), (\underline{3}, \sigma\rho), (\underline{0}, \rho^2), (\underline{0}, \sigma\rho^2) \right\}$$

$$\mathbf{B}_{447} = \left\{ (\underline{2}, 1), (\underline{3}, 1), (\underline{1}, \sigma), (\underline{3}, \sigma), (\underline{1}, \rho), (\underline{1}, \sigma\rho), (\underline{1}, \rho^2), (\underline{2}, \rho^2) \right\}$$

$$\mathbf{B}_{448} = \left\{ (\underline{2}, 1), (\underline{3}, 1), (\underline{1}, \sigma), (\underline{3}, \sigma), (\underline{2}, \rho), (\underline{0}, \sigma\rho), (\underline{1}, \sigma\rho^2), (\underline{3}, \sigma\rho^2) \right\}$$

$$\mathbf{B}_{449} = \left\{ (\underline{2}, 1), (\underline{3}, 1), (\underline{1}, \sigma), (\underline{3}, \sigma), (\underline{3}, \rho), (\underline{2}, \sigma\rho), (\underline{3}, \rho^2), (\underline{2}, \sigma\rho^2) \right\}$$

$$\mathbf{B}_{450} = \left\{ (\underline{2}, 1), (\underline{3}, 1), (\underline{2}, \sigma), (\underline{3}, \sigma), (\underline{0}, \rho), (\underline{1}, \rho), (\underline{3}, \rho), (\underline{0}, \sigma\rho) \right\}$$

$$\mathbf{B}_{451} = \left\{ (\underline{2}, 1), (\underline{3}, 1), (\underline{2}, \sigma), (\underline{3}, \sigma), (\underline{2}, \rho), (\underline{1}, \sigma\rho), (\underline{2}, \sigma\rho), (\underline{3}, \sigma\rho) \right\}$$

$$\mathbf{B}_{452} = \left\{ (\underline{2}, 1), (\underline{3}, 1), (\underline{2}, \sigma), (\underline{3}, \sigma), (\underline{0}, \rho^2), (\underline{2}, \rho^2), (\underline{2}, \sigma\rho^2), (\underline{3}, \sigma\rho^2) \right\}$$

$$\mathbf{B}_{453} = \left\{ (\underline{2}, 1), (\underline{3}, 1), (\underline{2}, \sigma), (\underline{3}, \sigma), (\underline{1}, \rho^2), (\underline{3}, \rho^2), (\underline{0}, \sigma\rho^2), (\underline{1}, \sigma\rho^2) \right\}$$

$$\mathbf{B}_{454} = \left\{ (\underline{2}, 1), (\underline{3}, 1), (\underline{0}, \rho), (\underline{1}, \rho), (\underline{2}, \rho), (\underline{1}, \sigma\rho), (\underline{0}, \rho^2), (\underline{3}, \sigma\rho^2) \right\}$$

$$\mathbf{B}_{455} = \left\{ (\underline{2}, 1), (\underline{3}, 1), (\underline{0}, \rho), (\underline{1}, \rho), (\underline{2}, \sigma\rho), (\underline{3}, \sigma\rho), (\underline{2}, \rho^2), (\underline{2}, \sigma\rho^2) \right\}$$

$$\mathbf{B}_{456} = \left\{ (\underline{2}, 1), (\underline{3}, 1), (\underline{0}, \rho), (\underline{2}, \rho), (\underline{3}, \rho), (\underline{2}, \sigma\rho), (\underline{0}, \sigma\rho^2), (\underline{1}, \sigma\rho^2) \right\}$$

$$\mathbf{B}_{457} = \left\{ (\underline{2}, 1), (\underline{3}, 1), (\underline{0}, \rho), (\underline{3}, \rho), (\underline{1}, \sigma\rho), (\underline{3}, \sigma\rho), (\underline{1}, \rho^2), (\underline{3}, \rho^2) \right\}$$

$$\mathbf{B}_{458} = \left\{ (\underline{2}, 1), (\underline{3}, 1), (\underline{0}, \rho), (\underline{0}, \sigma\rho), (\underline{0}, \rho^2), (\underline{1}, \rho^2), (\underline{2}, \rho^2), (\underline{1}, \sigma\rho^2) \right\}$$

$$\mathbf{B}_{459} = \left\{ (\underline{2}, 1), (\underline{3}, 1), (\underline{0}, \rho), (\underline{0}, \sigma\rho), (\underline{3}, \rho^2), (\underline{0}, \sigma\rho^2), (\underline{2}, \sigma\rho^2), (\underline{3}, \sigma\rho^2) \right\}$$

$$\mathbf{B}_{460} = \left\{ (\underline{2}, 1), (\underline{3}, 1), (\underline{1}, \rho), (\underline{2}, \rho), (\underline{0}, \sigma\rho), (\underline{2}, \sigma\rho), (\underline{1}, \rho^2), (\underline{3}, \rho^2) \right\}$$

$$\mathbf{B}_{461} = \left\{ (\underline{2}, 1), (\underline{3}, 1), (\underline{1}, \rho), (\underline{3}, \rho), (\underline{0}, \rho^2), (\underline{2}, \rho^2), (\underline{3}, \rho^2), (\underline{0}, \sigma\rho^2) \right\}$$

$$\mathbf{B}_{462} = \left\{ (\underline{2}, 1), (\underline{3}, 1), (\underline{1}, \rho), (\underline{3}, \rho), (\underline{1}, \rho^2), (\underline{1}, \sigma\rho^2), (\underline{2}, \sigma\rho^2), (\underline{3}, \sigma\rho^2) \right\}$$

$$\mathbf{B}_{463} = \left\{ (\underline{2}, 1), (\underline{3}, 1), (\underline{1}, \rho), (\underline{0}, \sigma\rho), (\underline{1}, \sigma\rho), (\underline{3}, \sigma\rho), (\underline{0}, \sigma\rho^2), (\underline{1}, \sigma\rho^2) \right\}$$

$$\mathbf{B}_{464} = \left\{ (\underline{2}, 1), (\underline{3}, 1), (\underline{2}, \rho), (\underline{3}, \rho), (\underline{0}, \sigma\rho), (\underline{1}, \sigma\rho), (\underline{2}, \rho^2), (\underline{2}, \sigma\rho^2) \right\}$$

$$\mathbf{B}_{465} = \left\{ (\underline{2}, 1), (\underline{3}, 1), (\underline{2}, \rho), (\underline{3}, \sigma\rho), (\underline{0}, \rho^2), (\underline{3}, \rho^2), (\underline{1}, \sigma\rho^2), (\underline{2}, \sigma\rho^2) \right\}$$

$$\mathbf{B}_{466} = \left\{ (\underline{2}, 1), (\underline{3}, 1), (\underline{2}, \rho), (\underline{3}, \sigma\rho), (\underline{1}, \rho^2), (\underline{2}, \rho^2), (\underline{0}, \sigma\rho^2), (\underline{3}, \sigma\rho^2) \right\}$$

$$\mathbf{B}_{467} = \left\{ (\underline{2}, 1), (\underline{3}, 1), (\underline{3}, \rho), (\underline{0}, \sigma\rho), (\underline{2}, \sigma\rho), (\underline{3}, \sigma\rho), (\underline{0}, \rho^2), (\underline{3}, \sigma\rho^2) \right\}$$

$$\mathbf{B}_{468} = \left\{ (\underline{2}, 1), (\underline{3}, 1), (\underline{1}, \sigma\rho), (\underline{2}, \sigma\rho), (\underline{0}, \rho^2), (\underline{1}, \rho^2), (\underline{0}, \sigma\rho^2), (\underline{2}, \sigma\rho^2) \right\}$$

$$\mathbf{B}_{469} = \left\{ (\underline{2}, 1), (\underline{3}, 1), (\underline{1}, \sigma\rho), (\underline{2}, \sigma\rho), (\underline{2}, \rho^2), (\underline{3}, \rho^2), (\underline{1}, \sigma\rho^2), (\underline{3}, \sigma\rho^2) \right\}$$

$$\mathbf{B}_{470} = \left\{ (\underline{2}, 1), (\underline{0}, \sigma), (\underline{1}, \sigma), (\underline{2}, \sigma), (\underline{0}, \rho), (\underline{1}, \rho), (\underline{0}, \rho^2), (\underline{2}, \rho^2) \right\}$$

$$\mathbf{B}_{471} = \left\{ (\underline{2}, 1), (\underline{0}, \sigma), (\underline{1}, \sigma), (\underline{2}, \sigma), (\underline{2}, \rho), (\underline{2}, \sigma\rho), (\underline{3}, \rho^2), (\underline{1}, \sigma\rho^2) \right\}$$

$$\mathbf{B}_{472} = \left\{ (\underline{2}, 1), (\underline{0}, \sigma), (\underline{1}, \sigma), (\underline{2}, \sigma), (\underline{3}, \rho), (\underline{0}, \sigma\rho), (\underline{2}, \sigma\rho^2), (\underline{3}, \sigma\rho^2) \right\}$$

$$\mathbf{B}_{473} = \left\{ (\underline{2}, 1), (\underline{0}, \sigma), (\underline{1}, \sigma), (\underline{2}, \sigma), (\underline{1}, \sigma\rho), (\underline{3}, \sigma\rho), (\underline{1}, \rho^2), (\underline{0}, \sigma\rho^2) \right\}$$

$$\mathbf{B}_{474} = \left\{ (\underline{2}, 1), (\underline{0}, \sigma), (\underline{1}, \sigma), (\underline{3}, \sigma), (\underline{0}, \rho), (\underline{3}, \rho), (\underline{1}, \rho^2), (\underline{1}, \sigma\rho^2) \right\}$$

$$\mathbf{B}_{475} = \left\{ (\underline{2}, 1), (\underline{0}, \sigma), (\underline{1}, \sigma), (\underline{3}, \sigma), (\underline{1}, \rho), (\underline{0}, \sigma\rho), (\underline{3}, \rho^2), (\underline{0}, \sigma\rho^2) \right\}$$

$$\mathbf{B}_{476} = \left\{ (\underline{2}, 1), (\underline{0}, \sigma), (\underline{1}, \sigma), (\underline{3}, \sigma), (\underline{2}, \rho), (\underline{1}, \sigma\rho), (\underline{0}, \rho^2), (\underline{2}, \sigma\rho^2) \right\}$$

$$\mathbf{B}_{477} = \left\{ (\underline{2}, 1), (\underline{0}, \sigma), (\underline{1}, \sigma), (\underline{3}, \sigma), (\underline{2}, \sigma\rho), (\underline{3}, \sigma\rho), (\underline{2}, \rho^2), (\underline{3}, \sigma\rho^2) \right\}$$

$$\mathbf{B}_{478} = \left\{ (\underline{2}, 1), (\underline{0}, \sigma), (\underline{2}, \sigma), (\underline{3}, \sigma), (\underline{0}, \rho), (\underline{2}, \sigma\rho), (\underline{0}, \sigma\rho^2), (\underline{2}, \sigma\rho^2) \right\}$$

$$\mathbf{B}_{479} = \left\{ (\underline{2}, 1), (\underline{0}, \sigma), (\underline{2}, \sigma), (\underline{3}, \sigma), (\underline{1}, \rho), (\underline{2}, \rho), (\underline{1}, \rho^2), (\underline{3}, \sigma\rho^2) \right\}$$

$$\mathbf{B}_{480} = \left\{ (\underline{2}, 1), (\underline{0}, \sigma), (\underline{2}, \sigma), (\underline{3}, \sigma), (\underline{3}, \rho), (\underline{3}, \sigma\rho), (\underline{0}, \rho^2), (\underline{3}, \rho^2) \right\}$$

$$\mathbf{B}_{481} = \left\{ (\underline{2}, 1), (\underline{0}, \sigma), (\underline{2}, \sigma), (\underline{3}, \sigma), (\underline{0}, \sigma\rho), (\underline{1}, \sigma\rho), (\underline{2}, \rho^2), (\underline{1}, \sigma\rho^2) \right\}$$

$$\mathbf{B}_{482} = \left\{ (\underline{2}, 1), (\underline{0}, \sigma), (\underline{0}, \rho), (\underline{1}, \rho), (\underline{2}, \rho), (\underline{0}, \sigma\rho), (\underline{1}, \sigma\rho^2), (\underline{2}, \sigma\rho^2) \right\}$$

$$\mathbf{B}_{483} = \left\{ (\underline{2}, 1), (\underline{0}, \sigma), (\underline{0}, \rho), (\underline{1}, \rho), (\underline{3}, \rho), (\underline{2}, \sigma\rho), (\underline{3}, \rho^2), (\underline{3}, \sigma\rho^2) \right\}$$

$$\mathbf{B}_{484} = \left\{ (\underline{2}, 1), (\underline{0}, \sigma), (\underline{0}, \rho), (\underline{2}, \rho), (\underline{2}, \sigma\rho), (\underline{3}, \sigma\rho), (\underline{0}, \rho^2), (\underline{1}, \rho^2) \right\}$$

$$\mathbf{B}_{485} = \left\{ (\underline{2}, 1), (\underline{0}, \sigma), (\underline{0}, \rho), (\underline{3}, \rho), (\underline{0}, \sigma\rho), (\underline{3}, \sigma\rho), (\underline{2}, \rho^2), (\underline{0}, \sigma\rho^2) \right\}$$

$$\mathbf{B}_{486} = \left\{ (\underline{2}, 1), (\underline{0}, \sigma), (\underline{0}, \rho), (\underline{1}, \sigma\rho), (\underline{0}, \rho^2), (\underline{3}, \rho^2), (\underline{0}, \sigma\rho^2), (\underline{1}, \sigma\rho^2) \right\}$$

$$\mathbf{B}_{487} = \left\{ (\underline{2}, 1), (\underline{0}, \sigma), (\underline{0}, \rho), (\underline{1}, \sigma\rho), (\underline{1}, \rho^2), (\underline{2}, \rho^2), (\underline{2}, \sigma\rho^2), (\underline{3}, \sigma\rho^2) \right\}$$

$$\mathbf{B}_{488} = \left\{ (\underline{2}, 1), (\underline{0}, \sigma), (\underline{1}, \rho), (\underline{2}, \rho), (\underline{1}, \sigma\rho), (\underline{2}, \sigma\rho), (\underline{2}, \rho^2), (\underline{0}, \sigma\rho^2) \right\}$$

$$\mathbf{B}_{489} = \left\{ (\underline{2}, 1), (\underline{0}, \sigma), (\underline{1}, \rho), (\underline{3}, \rho), (\underline{0}, \sigma\rho), (\underline{1}, \sigma\rho), (\underline{0}, \rho^2), (\underline{1}, \rho^2) \right\}$$

$$\mathbf{B}_{490} = \left\{ (\underline{2}, 1), (\underline{0}, \sigma), (\underline{1}, \rho), (\underline{3}, \sigma\rho), (\underline{0}, \rho^2), (\underline{0}, \sigma\rho^2), (\underline{2}, \sigma\rho^2), (\underline{3}, \sigma\rho^2) \right\}$$

$$\mathbf{B}_{491} = \left\{ (\underline{2}, 1), (\underline{0}, \sigma), (\underline{1}, \rho), (\underline{3}, \sigma\rho), (\underline{1}, \rho^2), (\underline{2}, \rho^2), (\underline{3}, \rho^2), (\underline{1}, \sigma\rho^2) \right\}$$

$$\mathbf{B}_{492} = \left\{ (\underline{2}, 1), (\underline{0}, \sigma), (\underline{2}, \rho), (\underline{3}, \rho), (\underline{0}, \rho^2), (\underline{2}, \rho^2), (\underline{1}, \sigma\rho^2), (\underline{3}, \sigma\rho^2) \right\}$$

$$\mathbf{B}_{493} = \left\{ (\underline{2}, 1), (\underline{0}, \sigma), (\underline{2}, \rho), (\underline{3}, \rho), (\underline{1}, \rho^2), (\underline{3}, \rho^2), (\underline{0}, \sigma\rho^2), (\underline{2}, \sigma\rho^2) \right\}$$

$$\mathbf{B}_{494} = \left\{ (\underline{2}, 1), (\underline{0}, \sigma), (\underline{2}, \rho), (\underline{0}, \sigma\rho), (\underline{1}, \sigma\rho), (\underline{3}, \sigma\rho), (\underline{3}, \rho^2), (\underline{3}, \sigma\rho^2) \right\}$$

$$\mathbf{B}_{495} = \left\{ (\underline{2}, 1), (\underline{0}, \sigma), (\underline{3}, \rho), (\underline{1}, \sigma\rho), (\underline{2}, \sigma\rho), (\underline{3}, \sigma\rho), (\underline{1}, \sigma\rho^2), (\underline{2}, \sigma\rho^2) \right\}$$

$$\mathbf{B}_{496} = \left\{ (\underline{2}, 1), (\underline{0}, \sigma), (\underline{0}, \sigma\rho), (\underline{2}, \sigma\rho), (\underline{0}, \rho^2), (\underline{2}, \rho^2), (\underline{3}, \rho^2), (\underline{2}, \sigma\rho^2) \right\}$$

$$\mathbf{B}_{497} = \left\{ (\underline{2}, 1), (\underline{0}, \sigma), (\underline{0}, \sigma\rho), (\underline{2}, \sigma\rho), (\underline{1}, \rho^2), (\underline{0}, \sigma\rho^2), (\underline{1}, \sigma\rho^2), (\underline{3}, \sigma\rho^2) \right\}$$

$$\mathbf{B}_{498} = \left\{ (\underline{2}, 1), (\underline{1}, \sigma), (\underline{2}, \sigma), (\underline{3}, \sigma), (\underline{0}, \rho), (\underline{1}, \sigma\rho), (\underline{3}, \rho^2), (\underline{3}, \sigma\rho^2) \right\}$$

$$\mathbf{B}_{499} = \left\{ (\underline{2}, 1), (\underline{1}, \sigma), (\underline{2}, \sigma), (\underline{3}, \sigma), (\underline{1}, \rho), (\underline{3}, \sigma\rho), (\underline{1}, \sigma\rho^2), (\underline{2}, \sigma\rho^2) \right\}$$

$$\mathbf{B}_{500} = \left\{ (\underline{2}, 1), (\underline{1}, \sigma), (\underline{2}, \sigma), (\underline{3}, \sigma), (\underline{2}, \rho), (\underline{3}, \rho), (\underline{2}, \rho^2), (\underline{0}, \sigma\rho^2) \right\}$$

$$\mathbf{B}_{501} = \left\{ (\underline{2}, 1), (\underline{1}, \sigma), (\underline{2}, \sigma), (\underline{3}, \sigma), (\underline{0}, \sigma\rho), (\underline{2}, \sigma\rho), (\underline{0}, \rho^2), (\underline{1}, \rho^2) \right\}$$

$$\mathbf{B}_{502} = \left\{ (\underline{2}, 1), (\underline{1}, \sigma), (\underline{0}, \rho), (\underline{1}, \rho), (\underline{3}, \rho), (\underline{1}, \sigma\rho), (\underline{0}, \sigma\rho^2), (\underline{2}, \sigma\rho^2) \right\}$$

$$\mathbf{B}_{503} = \left\{ (\underline{2}, 1), (\underline{1}, \sigma), (\underline{0}, \rho), (\underline{1}, \rho), (\underline{0}, \sigma\rho), (\underline{3}, \sigma\rho), (\underline{1}, \rho^2), (\underline{3}, \sigma\rho^2) \right\}$$

$$\mathbf{B}_{504} = \left\{ (\underline{2}, 1), (\underline{1}, \sigma), (\underline{0}, \rho), (\underline{2}, \rho), (\underline{3}, \rho), (\underline{0}, \sigma\rho), (\underline{0}, \rho^2), (\underline{3}, \rho^2) \right\}$$

$$\mathbf{B}_{505} = \left\{ (\underline{2}, 1), (\underline{1}, \sigma), (\underline{0}, \rho), (\underline{2}, \rho), (\underline{1}, \sigma\rho), (\underline{3}, \sigma\rho), (\underline{2}, \rho^2), (\underline{1}, \sigma\rho^2) \right\}$$

$$\mathbf{B}_{506} = \left\{ (\underline{2}, 1), (\underline{1}, \sigma), (\underline{0}, \rho), (\underline{2}, \sigma\rho), (\underline{0}, \rho^2), (\underline{1}, \sigma\rho^2), (\underline{2}, \sigma\rho^2), (\underline{3}, \sigma\rho^2) \right\}$$

$$\mathbf{B}_{507} = \left\{ (\underline{2}, 1), (\underline{1}, \sigma), (\underline{0}, \rho), (\underline{2}, \sigma\rho), (\underline{1}, \rho^2), (\underline{2}, \rho^2), (\underline{3}, \rho^2), (\underline{0}, \sigma\rho^2) \right\}$$

$$\mathbf{B}_{508} = \left\{ (\underline{2}, 1), (\underline{1}, \sigma), (\underline{1}, \rho), (\underline{2}, \rho), (\underline{0}, \rho^2), (\underline{1}, \rho^2), (\underline{0}, \sigma\rho^2), (\underline{1}, \sigma\rho^2) \right\}$$

$$\mathbf{B}_{509} = \left\{ (\underline{2}, 1), (\underline{1}, \sigma), (\underline{1}, \rho), (\underline{2}, \rho), (\underline{2}, \rho^2), (\underline{3}, \rho^2), (\underline{2}, \sigma\rho^2), (\underline{3}, \sigma\rho^2) \right\}$$

$$\mathbf{B}_{510} = \left\{ (\underline{2}, 1), (\underline{1}, \sigma), (\underline{1}, \rho), (\underline{3}, \rho), (\underline{0}, \sigma\rho), (\underline{2}, \sigma\rho), (\underline{2}, \rho^2), (\underline{1}, \sigma\rho^2) \right\}$$

$$\mathbf{B}_{511} = \left\{ (\underline{2}, 1), (\underline{1}, \sigma), (\underline{1}, \rho), (\underline{1}, \sigma\rho), (\underline{2}, \sigma\rho), (\underline{3}, \sigma\rho), (\underline{0}, \rho^2), (\underline{3}, \rho^2) \right\}$$

$$\mathbf{B}_{512} = \left\{ (\underline{2}, 1), (\underline{1}, \sigma), (\underline{2}, \rho), (\underline{3}, \rho), (\underline{1}, \sigma\rho), (\underline{2}, \sigma\rho), (\underline{1}, \rho^2), (\underline{3}, \sigma\rho^2) \right\}$$

$$\mathbf{B}_{513} = \left\{ (\underline{2}, 1), (\underline{1}, \sigma), (\underline{2}, \rho), (\underline{0}, \sigma\rho), (\underline{2}, \sigma\rho), (\underline{3}, \sigma\rho), (\underline{0}, \sigma\rho^2), (\underline{2}, \sigma\rho^2) \right\}$$

$$\mathbf{B}_{514} = \left\{ (\underline{2}, 1), (\underline{1}, \sigma), (\underline{3}, \rho), (\underline{3}, \sigma\rho), (\underline{0}, \rho^2), (\underline{1}, \rho^2), (\underline{2}, \rho^2), (\underline{2}, \sigma\rho^2) \right\}$$

$$\mathbf{B}_{515} = \left\{ (\underline{2}, 1), (\underline{1}, \sigma), (\underline{3}, \rho), (\underline{3}, \sigma\rho), (\underline{3}, \rho^2), (\underline{0}, \sigma\rho^2), (\underline{1}, \sigma\rho^2), (\underline{3}, \sigma\rho^2) \right\}$$

$$\mathbf{B}_{516} = \left\{ (\underline{2}, 1), (\underline{1}, \sigma), (\underline{0}, \sigma\rho), (\underline{1}, \sigma\rho), (\underline{0}, \rho^2), (\underline{2}, \rho^2), (\underline{0}, \sigma\rho^2), (\underline{3}, \sigma\rho^2) \right\}$$

$$\mathbf{B}_{517} = \left\{ (\underline{2}, 1), (\underline{1}, \sigma), (\underline{0}, \sigma\rho), (\underline{1}, \sigma\rho), (\underline{1}, \rho^2), (\underline{3}, \rho^2), (\underline{1}, \sigma\rho^2), (\underline{2}, \sigma\rho^2) \right\}$$

$$\mathbf{B}_{518} = \left\{ (\underline{2}, 1), (\underline{2}, \sigma), (\underline{0}, \rho), (\underline{1}, \rho), (\underline{2}, \rho), (\underline{3}, \sigma\rho), (\underline{3}, \rho^2), (\underline{0}, \sigma\rho^2) \right\}$$

$$\mathbf{B}_{519} = \left\{ (\underline{2}, 1), (\underline{2}, \sigma), (\underline{0}, \rho), (\underline{1}, \rho), (\underline{1}, \sigma\rho), (\underline{2}, \sigma\rho), (\underline{1}, \rho^2), (\underline{1}, \sigma\rho^2) \right\}$$

$$\mathbf{B}_{520} = \left\{ (\underline{2}, 1), (\underline{2}, \sigma), (\underline{0}, \rho), (\underline{2}, \rho), (\underline{0}, \sigma\rho), (\underline{2}, \sigma\rho), (\underline{2}, \rho^2), (\underline{3}, \sigma\rho^2) \right\}$$

$$\mathbf{B}_{521} = \left\{ (\underline{2}, 1), (\underline{2}, \sigma), (\underline{0}, \rho), (\underline{3}, \rho), (\underline{0}, \rho^2), (\underline{1}, \rho^2), (\underline{0}, \sigma\rho^2), (\underline{3}, \sigma\rho^2) \right\}$$

$$\mathbf{B}_{522} = \left\{ (\underline{2}, 1), (\underline{2}, \sigma), (\underline{0}, \rho), (\underline{3}, \rho), (\underline{2}, \rho^2), (\underline{3}, \rho^2), (\underline{1}, \sigma\rho^2), (\underline{2}, \sigma\rho^2) \right\}$$

$$\mathbf{B}_{523} = \left\{ (\underline{2}, 1), (\underline{2}, \sigma), (\underline{0}, \rho), (\underline{0}, \sigma\rho), (\underline{1}, \sigma\rho), (\underline{3}, \sigma\rho), (\underline{0}, \rho^2), (\underline{2}, \sigma\rho^2) \right\}$$

$$\mathbf{B}_{524} = \left\{ (\underline{2}, 1), (\underline{2}, \sigma), (\underline{1}, \rho), (\underline{2}, \rho), (\underline{3}, \rho), (\underline{2}, \sigma\rho), (\underline{0}, \rho^2), (\underline{2}, \sigma\rho^2) \right\}$$

$$\mathbf{B}_{525} = \left\{ (\underline{2}, 1), (\underline{2}, \sigma), (\underline{1}, \rho), (\underline{3}, \rho), (\underline{1}, \sigma\rho), (\underline{3}, \sigma\rho), (\underline{2}, \rho^2), (\underline{3}, \sigma\rho^2) \right\}$$

$$\mathbf{B}_{526} = \left\{ (\underline{2}, 1), (\underline{2}, \sigma), (\underline{1}, \rho), (\underline{0}, \sigma\rho), (\underline{0}, \rho^2), (\underline{3}, \rho^2), (\underline{1}, \sigma\rho^2), (\underline{3}, \sigma\rho^2) \right\}$$

$$\mathbf{B}_{527} = \left\{ (\underline{2}, 1), (\underline{2}, \sigma), (\underline{1}, \rho), (\underline{0}, \sigma\rho), (\underline{1}, \rho^2), (\underline{2}, \rho^2), (\underline{0}, \sigma\rho^2), (\underline{2}, \sigma\rho^2) \right\}$$

$$\mathbf{B}_{528} = \left\{ (\underline{2}, 1), (\underline{2}, \sigma), (\underline{2}, \rho), (\underline{3}, \rho), (\underline{0}, \sigma\rho), (\underline{3}, \sigma\rho), (\underline{1}, \rho^2), (\underline{1}, \sigma\rho^2) \right\}$$

$$\mathbf{B}_{529} = \left\{ (\underline{2}, 1), (\underline{2}, \sigma), (\underline{2}, \rho), (\underline{1}, \sigma\rho), (\underline{0}, \rho^2), (\underline{1}, \rho^2), (\underline{2}, \rho^2), (\underline{3}, \rho^2) \right\}$$

$$\mathbf{B}_{530} = \left\{ (\underline{2}, 1), (\underline{2}, \sigma), (\underline{2}, \rho), (\underline{1}, \sigma\rho), (\underline{0}, \sigma\rho^2), (\underline{1}, \sigma\rho^2), (\underline{2}, \sigma\rho^2), (\underline{3}, \sigma\rho^2) \right\}$$

$$\mathbf{B}_{531} = \left\{ (\underline{2}, 1), (\underline{2}, \sigma), (\underline{3}, \rho), (\underline{0}, \sigma\rho), (\underline{1}, \sigma\rho), (\underline{2}, \sigma\rho), (\underline{3}, \rho^2), (\underline{0}, \sigma\rho^2) \right\}$$

$$\mathbf{B}_{532} = \left\{ (\underline{2}, 1), (\underline{2}, \sigma), (\underline{2}, \sigma\rho), (\underline{3}, \sigma\rho), (\underline{0}, \rho^2), (\underline{2}, \rho^2), (\underline{0}, \sigma\rho^2), (\underline{1}, \sigma\rho^2) \right\}$$

$$\mathbf{B}_{533} = \left\{ (\underline{2}, 1), (\underline{2}, \sigma), (\underline{2}, \sigma\rho), (\underline{3}, \sigma\rho), (\underline{1}, \rho^2), (\underline{3}, \rho^2), (\underline{2}, \sigma\rho^2), (\underline{3}, \sigma\rho^2) \right\}$$

$\mathbf{B}_{534} = \left\{ (\underline{2}, 1), (\underline{3}, \sigma), (\underline{0}, \rho), (\underline{1}, \rho), (\underline{0}, \rho^2), (\underline{1}, \rho^2), (\underline{3}, \rho^2), (\underline{2}, \sigma\rho^2) \right\}$

$\mathbf{B}_{535} = \left\{ (\underline{2}, 1), (\underline{3}, \sigma), (\underline{0}, \rho), (\underline{1}, \rho), (\underline{2}, \rho^2), (\underline{0}, \sigma\rho^2), (\underline{1}, \sigma\rho^2), (\underline{3}, \sigma\rho^2) \right\}$

$\mathbf{B}_{536} = \left\{ (\underline{2}, 1), (\underline{3}, \sigma), (\underline{0}, \rho), (\underline{2}, \rho), (\underline{3}, \rho), (\underline{3}, \sigma\rho), (\underline{2}, \sigma\rho^2), (\underline{3}, \sigma\rho^2) \right\}$

$\mathbf{B}_{537} = \left\{ (\underline{2}, 1), (\underline{3}, \sigma), (\underline{0}, \rho), (\underline{2}, \rho), (\underline{0}, \sigma\rho), (\underline{1}, \sigma\rho), (\underline{1}, \rho^2), (\underline{0}, \sigma\rho^2) \right\}$

$\mathbf{B}_{538} = \left\{ (\underline{2}, 1), (\underline{3}, \sigma), (\underline{0}, \rho), (\underline{3}, \rho), (\underline{1}, \sigma\rho), (\underline{2}, \sigma\rho), (\underline{0}, \rho^2), (\underline{2}, \rho^2) \right\}$

$\mathbf{B}_{539} = \left\{ (\underline{2}, 1), (\underline{3}, \sigma), (\underline{0}, \rho), (\underline{0}, \sigma\rho), (\underline{2}, \sigma\rho), (\underline{3}, \sigma\rho), (\underline{3}, \rho^2), (\underline{1}, \sigma\rho^2) \right\}$

$\mathbf{B}_{540} = \left\{ (\underline{2}, 1), (\underline{3}, \sigma), (\underline{1}, \rho), (\underline{2}, \rho), (\underline{3}, \rho), (\underline{1}, \sigma\rho), (\underline{3}, \rho^2), (\underline{1}, \sigma\rho^2) \right\}$

$\mathbf{B}_{541} = \left\{ (\underline{2}, 1), (\underline{3}, \sigma), (\underline{1}, \rho), (\underline{2}, \rho), (\underline{0}, \sigma\rho), (\underline{3}, \sigma\rho), (\underline{0}, \rho^2), (\underline{2}, \rho^2) \right\}$

$\mathbf{B}_{542} = \left\{ (\underline{2}, 1), (\underline{3}, \sigma), (\underline{1}, \rho), (\underline{3}, \rho), (\underline{2}, \sigma\rho), (\underline{3}, \sigma\rho), (\underline{1}, \rho^2), (\underline{0}, \sigma\rho^2) \right\}$

$\mathbf{B}_{543} = \left\{ (\underline{2}, 1), (\underline{3}, \sigma), (\underline{1}, \rho), (\underline{0}, \sigma\rho), (\underline{1}, \sigma\rho), (\underline{2}, \sigma\rho), (\underline{2}, \sigma\rho^2), (\underline{3}, \sigma\rho^2) \right\}$

$\mathbf{B}_{544} = \left\{ (\underline{2}, 1), (\underline{3}, \sigma), (\underline{2}, \rho), (\underline{2}, \sigma\rho), (\underline{0}, \rho^2), (\underline{3}, \rho^2), (\underline{0}, \sigma\rho^2), (\underline{3}, \sigma\rho^2) \right\}$

$\mathbf{B}_{545} = \left\{ (\underline{2}, 1), (\underline{3}, \sigma), (\underline{2}, \rho), (\underline{2}, \sigma\rho), (\underline{1}, \rho^2), (\underline{2}, \rho^2), (\underline{1}, \sigma\rho^2), (\underline{2}, \sigma\rho^2) \right\}$

$\mathbf{B}_{546} = \left\{ (\underline{2}, 1), (\underline{3}, \sigma), (\underline{3}, \rho), (\underline{0}, \sigma\rho), (\underline{0}, \rho^2), (\underline{0}, \sigma\rho^2), (\underline{1}, \sigma\rho^2), (\underline{2}, \sigma\rho^2) \right\}$

$\mathbf{B}_{547} = \left\{ (\underline{2}, 1), (\underline{3}, \sigma), (\underline{3}, \rho), (\underline{0}, \sigma\rho), (\underline{1}, \rho^2), (\underline{2}, \rho^2), (\underline{3}, \rho^2), (\underline{3}, \sigma\rho^2) \right\}$

$\mathbf{B}_{548} = \left\{ (\underline{2}, 1), (\underline{3}, \sigma), (\underline{1}, \sigma\rho), (\underline{3}, \sigma\rho), (\underline{0}, \rho^2), (\underline{1}, \rho^2), (\underline{1}, \sigma\rho^2), (\underline{3}, \sigma\rho^2) \right\}$

$\mathbf{B}_{549} = \left\{ (\underline{2}, 1), (\underline{3}, \sigma), (\underline{1}, \sigma\rho), (\underline{3}, \sigma\rho), (\underline{2}, \rho^2), (\underline{3}, \rho^2), (\underline{0}, \sigma\rho^2), (\underline{2}, \sigma\rho^2) \right\}$

$\mathbf{B}_{550} = \left\{ (\underline{3}, 1), (\underline{0}, \sigma), (\underline{1}, \sigma), (\underline{2}, \sigma), (\underline{0}, \rho), (\underline{3}, \rho), (\underline{3}, \rho^2), (\underline{0}, \sigma\rho^2) \right\}$

$\mathbf{B}_{551} = \left\{ (\underline{3}, 1), (\underline{0}, \sigma), (\underline{1}, \sigma), (\underline{2}, \sigma), (\underline{1}, \rho), (\underline{0}, \sigma\rho), (\underline{1}, \rho^2), (\underline{1}, \sigma\rho^2) \right\}$

$\mathbf{B}_{552} = \left\{ (\underline{3}, 1), (\underline{0}, \sigma), (\underline{1}, \sigma), (\underline{2}, \sigma), (\underline{2}, \rho), (\underline{1}, \sigma\rho), (\underline{2}, \rho^2), (\underline{3}, \sigma\rho^2) \right\}$

$\mathbf{B}_{553} = \left\{ (\underline{3}, 1), (\underline{0}, \sigma), (\underline{1}, \sigma), (\underline{2}, \sigma), (\underline{2}, \sigma\rho), (\underline{3}, \sigma\rho), (\underline{0}, \rho^2), (\underline{2}, \sigma\rho^2) \right\}$

$\mathbf{B}_{554} = \left\{ (\underline{3}, 1), (\underline{0}, \sigma), (\underline{1}, \sigma), (\underline{3}, \sigma), (\underline{0}, \rho), (\underline{1}, \rho), (\underline{2}, \sigma\rho^2), (\underline{3}, \sigma\rho^2) \right\}$

$\mathbf{B}_{555} = \left\{ (\underline{3}, 1), (\underline{0}, \sigma), (\underline{1}, \sigma), (\underline{3}, \sigma), (\underline{2}, \rho), (\underline{2}, \sigma\rho), (\underline{1}, \rho^2), (\underline{0}, \sigma\rho^2) \right\}$

$$\mathbf{B}_{556} = \left\{ (\underline{3}, 1), (\underline{0}, \sigma), (\underline{1}, \sigma), (\underline{3}, \sigma), (\underline{3}, \rho), (\underline{0}, \sigma\rho), (\underline{0}, \rho^2), (\underline{2}, \rho^2) \right\}$$

$$\mathbf{B}_{557} = \left\{ (\underline{3}, 1), (\underline{0}, \sigma), (\underline{1}, \sigma), (\underline{3}, \sigma), (\underline{1}, \sigma\rho), (\underline{3}, \sigma\rho), (\underline{3}, \rho^2), (\underline{1}, \sigma\rho^2) \right\}$$

$$\mathbf{B}_{558} = \left\{ (\underline{3}, 1), (\underline{0}, \sigma), (\underline{2}, \sigma), (\underline{3}, \sigma), (\underline{0}, \rho), (\underline{1}, \sigma\rho), (\underline{0}, \rho^2), (\underline{1}, \rho^2) \right\}$$

$$\mathbf{B}_{559} = \left\{ (\underline{3}, 1), (\underline{0}, \sigma), (\underline{2}, \sigma), (\underline{3}, \sigma), (\underline{1}, \rho), (\underline{3}, \sigma\rho), (\underline{2}, \rho^2), (\underline{0}, \sigma\rho^2) \right\}$$

$$\mathbf{B}_{560} = \left\{ (\underline{3}, 1), (\underline{0}, \sigma), (\underline{2}, \sigma), (\underline{3}, \sigma), (\underline{2}, \rho), (\underline{3}, \rho), (\underline{1}, \sigma\rho^2), (\underline{2}, \sigma\rho^2) \right\}$$

$$\mathbf{B}_{561} = \left\{ (\underline{3}, 1), (\underline{0}, \sigma), (\underline{2}, \sigma), (\underline{3}, \sigma), (\underline{0}, \sigma\rho), (\underline{2}, \sigma\rho), (\underline{3}, \rho^2), (\underline{3}, \sigma\rho^2) \right\}$$

$$\mathbf{B}_{562} = \left\{ (\underline{3}, 1), (\underline{0}, \sigma), (\underline{0}, \rho), (\underline{1}, \rho), (\underline{3}, \rho), (\underline{1}, \sigma\rho), (\underline{2}, \rho^2), (\underline{1}, \sigma\rho^2) \right\}$$

$$\mathbf{B}_{563} = \left\{ (\underline{3}, 1), (\underline{0}, \sigma), (\underline{0}, \rho), (\underline{1}, \rho), (\underline{0}, \sigma\rho), (\underline{3}, \sigma\rho), (\underline{0}, \rho^2), (\underline{3}, \rho^2) \right\}$$

$$\mathbf{B}_{564} = \left\{ (\underline{3}, 1), (\underline{0}, \sigma), (\underline{0}, \rho), (\underline{2}, \rho), (\underline{3}, \rho), (\underline{0}, \sigma\rho), (\underline{1}, \rho^2), (\underline{3}, \sigma\rho^2) \right\}$$

$$\mathbf{B}_{565} = \left\{ (\underline{3}, 1), (\underline{0}, \sigma), (\underline{0}, \rho), (\underline{2}, \rho), (\underline{1}, \sigma\rho), (\underline{3}, \sigma\rho), (\underline{0}, \sigma\rho^2), (\underline{2}, \sigma\rho^2) \right\}$$

$$\mathbf{B}_{566} = \left\{ (\underline{3}, 1), (\underline{0}, \sigma), (\underline{0}, \rho), (\underline{2}, \sigma\rho), (\underline{0}, \rho^2), (\underline{2}, \rho^2), (\underline{0}, \sigma\rho^2), (\underline{3}, \sigma\rho^2) \right\}$$

$$\mathbf{B}_{567} = \left\{ (\underline{3}, 1), (\underline{0}, \sigma), (\underline{0}, \rho), (\underline{2}, \sigma\rho), (\underline{1}, \rho^2), (\underline{3}, \rho^2), (\underline{1}, \sigma\rho^2), (\underline{2}, \sigma\rho^2) \right\}$$

$$\mathbf{B}_{568} = \left\{ (\underline{3}, 1), (\underline{0}, \sigma), (\underline{1}, \rho), (\underline{2}, \rho), (\underline{0}, \rho^2), (\underline{1}, \rho^2), (\underline{2}, \rho^2), (\underline{2}, \sigma\rho^2) \right\}$$

$$\mathbf{B}_{569} = \left\{ (\underline{3}, 1), (\underline{0}, \sigma), (\underline{1}, \rho), (\underline{2}, \rho), (\underline{3}, \rho^2), (\underline{0}, \sigma\rho^2), (\underline{1}, \sigma\rho^2), (\underline{3}, \sigma\rho^2) \right\}$$

$$\mathbf{B}_{570} = \left\{ (\underline{3}, 1), (\underline{0}, \sigma), (\underline{1}, \rho), (\underline{3}, \rho), (\underline{0}, \sigma\rho), (\underline{2}, \sigma\rho), (\underline{0}, \sigma\rho^2), (\underline{2}, \sigma\rho^2) \right\}$$

$$\mathbf{B}_{571} = \left\{ (\underline{3}, 1), (\underline{0}, \sigma), (\underline{1}, \rho), (\underline{1}, \sigma\rho), (\underline{2}, \sigma\rho), (\underline{3}, \sigma\rho), (\underline{1}, \rho^2), (\underline{3}, \sigma\rho^2) \right\}$$

$$\mathbf{B}_{572} = \left\{ (\underline{3}, 1), (\underline{0}, \sigma), (\underline{2}, \rho), (\underline{3}, \rho), (\underline{1}, \sigma\rho), (\underline{2}, \sigma\rho), (\underline{0}, \rho^2), (\underline{3}, \rho^2) \right\}$$

$$\mathbf{B}_{573} = \left\{ (\underline{3}, 1), (\underline{0}, \sigma), (\underline{2}, \rho), (\underline{0}, \sigma\rho), (\underline{2}, \sigma\rho), (\underline{3}, \sigma\rho), (\underline{2}, \rho^2), (\underline{1}, \sigma\rho^2) \right\}$$

$$\mathbf{B}_{574} = \left\{ (\underline{3}, 1), (\underline{0}, \sigma), (\underline{3}, \rho), (\underline{3}, \sigma\rho), (\underline{0}, \rho^2), (\underline{1}, \rho^2), (\underline{0}, \sigma\rho^2), (\underline{1}, \sigma\rho^2) \right\}$$

$$\mathbf{B}_{575} = \left\{ (\underline{3}, 1), (\underline{0}, \sigma), (\underline{3}, \rho), (\underline{3}, \sigma\rho), (\underline{2}, \rho^2), (\underline{3}, \rho^2), (\underline{2}, \sigma\rho^2), (\underline{3}, \sigma\rho^2) \right\}$$

$$\mathbf{B}_{576} = \left\{ (\underline{3}, 1), (\underline{0}, \sigma), (\underline{0}, \sigma\rho), (\underline{1}, \sigma\rho), (\underline{0}, \rho^2), (\underline{1}, \sigma\rho^2), (\underline{2}, \sigma\rho^2), (\underline{3}, \sigma\rho^2) \right\}$$

$$\mathbf{B}_{577} = \left\{ (\underline{3}, 1), (\underline{0}, \sigma), (\underline{0}, \sigma\rho), (\underline{1}, \sigma\rho), (\underline{1}, \rho^2), (\underline{2}, \rho^2), (\underline{3}, \rho^2), (\underline{0}, \sigma\rho^2) \right\}$$

$$\mathbf{B}_{578} = \left\{ (\underline{3}, 1), (\underline{1}, \sigma), (\underline{2}, \sigma), (\underline{3}, \sigma), (\underline{0}, \rho), (\underline{2}, \sigma\rho), (\underline{2}, \rho^2), (\underline{1}, \sigma\rho^2) \right\}$$

$$\mathbf{B}_{579} = \left\{ (\underline{3}, 1), (\underline{1}, \sigma), (\underline{2}, \sigma), (\underline{3}, \sigma), (\underline{1}, \rho), (\underline{2}, \rho), (\underline{0}, \rho^2), (\underline{3}, \rho^2) \right\}$$

$$\mathbf{B}_{580} = \left\{ (\underline{3}, 1), (\underline{1}, \sigma), (\underline{2}, \sigma), (\underline{3}, \sigma), (\underline{3}, \rho), (\underline{3}, \sigma\rho), (\underline{1}, \rho^2), (\underline{3}, \sigma\rho^2) \right\}$$

$$\mathbf{B}_{581} = \left\{ (\underline{3}, 1), (\underline{1}, \sigma), (\underline{2}, \sigma), (\underline{3}, \sigma), (\underline{0}, \sigma\rho), (\underline{1}, \sigma\rho), (\underline{0}, \sigma\rho^2), (\underline{2}, \sigma\rho^2) \right\}$$

$$\mathbf{B}_{582} = \left\{ (\underline{3}, 1), (\underline{1}, \sigma), (\underline{0}, \rho), (\underline{1}, \rho), (\underline{2}, \rho), (\underline{0}, \sigma\rho), (\underline{2}, \rho^2), (\underline{0}, \sigma\rho^2) \right\}$$

$$\mathbf{B}_{583} = \left\{ (\underline{3}, 1), (\underline{1}, \sigma), (\underline{0}, \rho), (\underline{1}, \rho), (\underline{3}, \rho), (\underline{2}, \sigma\rho), (\underline{0}, \rho^2), (\underline{1}, \rho^2) \right\}$$

$$\mathbf{B}_{584} = \left\{ (\underline{3}, 1), (\underline{1}, \sigma), (\underline{0}, \rho), (\underline{2}, \rho), (\underline{2}, \sigma\rho), (\underline{3}, \sigma\rho), (\underline{3}, \rho^2), (\underline{3}, \sigma\rho^2) \right\}$$

$$\mathbf{B}_{585} = \left\{ (\underline{3}, 1), (\underline{1}, \sigma), (\underline{0}, \rho), (\underline{3}, \rho), (\underline{0}, \sigma\rho), (\underline{3}, \sigma\rho), (\underline{1}, \sigma\rho^2), (\underline{2}, \sigma\rho^2) \right\}$$

$$\mathbf{B}_{586} = \left\{ (\underline{3}, 1), (\underline{1}, \sigma), (\underline{0}, \rho), (\underline{1}, \sigma\rho), (\underline{0}, \rho^2), (\underline{2}, \rho^2), (\underline{3}, \rho^2), (\underline{2}, \sigma\rho^2) \right\}$$

$$\mathbf{B}_{587} = \left\{ (\underline{3}, 1), (\underline{1}, \sigma), (\underline{0}, \rho), (\underline{1}, \sigma\rho), (\underline{1}, \rho^2), (\underline{0}, \sigma\rho^2), (\underline{1}, \sigma\rho^2), (\underline{3}, \sigma\rho^2) \right\}$$

$$\mathbf{B}_{588} = \left\{ (\underline{3}, 1), (\underline{1}, \sigma), (\underline{1}, \rho), (\underline{2}, \rho), (\underline{1}, \sigma\rho), (\underline{2}, \sigma\rho), (\underline{1}, \sigma\rho^2), (\underline{2}, \sigma\rho^2) \right\}$$

$$\mathbf{B}_{589} = \left\{ (\underline{3}, 1), (\underline{1}, \sigma), (\underline{1}, \rho), (\underline{3}, \rho), (\underline{0}, \sigma\rho), (\underline{1}, \sigma\rho), (\underline{3}, \rho^2), (\underline{3}, \sigma\rho^2) \right\}$$

$$\mathbf{B}_{590} = \left\{ (\underline{3}, 1), (\underline{1}, \sigma), (\underline{1}, \rho), (\underline{3}, \sigma\rho), (\underline{0}, \rho^2), (\underline{2}, \rho^2), (\underline{1}, \sigma\rho^2), (\underline{3}, \sigma\rho^2) \right\}$$

$$\mathbf{B}_{591} = \left\{ (\underline{3}, 1), (\underline{1}, \sigma), (\underline{1}, \rho), (\underline{3}, \sigma\rho), (\underline{1}, \rho^2), (\underline{3}, \rho^2), (\underline{0}, \sigma\rho^2), (\underline{2}, \sigma\rho^2) \right\}$$

$$\mathbf{B}_{592} = \left\{ (\underline{3}, 1), (\underline{1}, \sigma), (\underline{2}, \rho), (\underline{3}, \rho), (\underline{0}, \rho^2), (\underline{0}, \sigma\rho^2), (\underline{2}, \sigma\rho^2), (\underline{3}, \sigma\rho^2) \right\}$$

$$\mathbf{B}_{593} = \left\{ (\underline{3}, 1), (\underline{1}, \sigma), (\underline{2}, \rho), (\underline{3}, \rho), (\underline{1}, \rho^2), (\underline{2}, \rho^2), (\underline{3}, \rho^2), (\underline{1}, \sigma\rho^2) \right\}$$

$$\mathbf{B}_{594} = \left\{ (\underline{3}, 1), (\underline{1}, \sigma), (\underline{2}, \rho), (\underline{0}, \sigma\rho), (\underline{1}, \sigma\rho), (\underline{3}, \sigma\rho), (\underline{0}, \rho^2), (\underline{1}, \rho^2) \right\}$$

$$\mathbf{B}_{595} = \left\{ (\underline{3}, 1), (\underline{1}, \sigma), (\underline{3}, \rho), (\underline{1}, \sigma\rho), (\underline{2}, \sigma\rho), (\underline{3}, \sigma\rho), (\underline{2}, \rho^2), (\underline{0}, \sigma\rho^2) \right\}$$

$$\mathbf{B}_{596} = \left\{ (\underline{3}, 1), (\underline{1}, \sigma), (\underline{0}, \sigma\rho), (\underline{2}, \sigma\rho), (\underline{0}, \rho^2), (\underline{3}, \rho^2), (\underline{0}, \sigma\rho^2), (\underline{1}, \sigma\rho^2) \right\}$$

$$\mathbf{B}_{597} = \left\{ (\underline{3}, 1), (\underline{1}, \sigma), (\underline{0}, \sigma\rho), (\underline{2}, \sigma\rho), (\underline{1}, \rho^2), (\underline{2}, \rho^2), (\underline{2}, \sigma\rho^2), (\underline{3}, \sigma\rho^2) \right\}$$

$$\mathbf{B}_{598} = \left\{ (\underline{3}, 1), (\underline{2}, \sigma), (\underline{0}, \rho), (\underline{1}, \rho), (\underline{0}, \rho^2), (\underline{0}, \sigma\rho^2), (\underline{1}, \sigma\rho^2), (\underline{2}, \sigma\rho^2) \right\}$$

$$\mathbf{B}_{599} = \left\{ (\underline{3}, 1), (\underline{2}, \sigma), (\underline{0}, \rho), (\underline{1}, \rho), (\underline{1}, \rho^2), (\underline{2}, \rho^2), (\underline{3}, \rho^2), (\underline{3}, \sigma\rho^2) \right\}$$

$\mathbf{B}_{600} = \left\{ (\underline{3}, 1), (\underline{2}, \sigma), (\underline{0}, \rho), (\underline{2}, \rho), (\underline{3}, \rho), (\underline{3}, \sigma\rho), (\underline{0}, \rho^2), (\underline{2}, \rho^2) \right\}$

$\mathbf{B}_{601} = \left\{ (\underline{3}, 1), (\underline{2}, \sigma), (\underline{0}, \rho), (\underline{2}, \rho), (\underline{0}, \sigma\rho), (\underline{1}, \sigma\rho), (\underline{3}, \rho^2), (\underline{1}, \sigma\rho^2) \right\}$

$\mathbf{B}_{602} = \left\{ (\underline{3}, 1), (\underline{2}, \sigma), (\underline{0}, \rho), (\underline{3}, \rho), (\underline{1}, \sigma\rho), (\underline{2}, \sigma\rho), (\underline{2}, \sigma\rho^2), (\underline{3}, \sigma\rho^2) \right\}$

$\mathbf{B}_{603} = \left\{ (\underline{3}, 1), (\underline{2}, \sigma), (\underline{0}, \rho), (\underline{0}, \sigma\rho), (\underline{2}, \sigma\rho), (\underline{3}, \sigma\rho), (\underline{1}, \rho^2), (\underline{0}, \sigma\rho^2) \right\}$

$\mathbf{B}_{604} = \left\{ (\underline{3}, 1), (\underline{2}, \sigma), (\underline{1}, \rho), (\underline{2}, \rho), (\underline{3}, \rho), (\underline{1}, \sigma\rho), (\underline{1}, \rho^2), (\underline{0}, \sigma\rho^2) \right\}$

$\mathbf{B}_{605} = \left\{ (\underline{3}, 1), (\underline{2}, \sigma), (\underline{1}, \rho), (\underline{2}, \rho), (\underline{0}, \sigma\rho), (\underline{3}, \sigma\rho), (\underline{2}, \sigma\rho^2), (\underline{3}, \sigma\rho^2) \right\}$

$\mathbf{B}_{606} = \left\{ (\underline{3}, 1), (\underline{2}, \sigma), (\underline{1}, \rho), (\underline{3}, \rho), (\underline{2}, \sigma\rho), (\underline{3}, \sigma\rho), (\underline{3}, \rho^2), (\underline{1}, \sigma\rho^2) \right\}$

$\mathbf{B}_{607} = \left\{ (\underline{3}, 1), (\underline{2}, \sigma), (\underline{1}, \rho), (\underline{0}, \sigma\rho), (\underline{1}, \sigma\rho), (\underline{2}, \sigma\rho), (\underline{0}, \rho^2), (\underline{2}, \rho^2) \right\}$

$\mathbf{B}_{608} = \left\{ (\underline{3}, 1), (\underline{2}, \sigma), (\underline{2}, \rho), (\underline{2}, \sigma\rho), (\underline{0}, \rho^2), (\underline{1}, \rho^2), (\underline{1}, \sigma\rho^2), (\underline{3}, \sigma\rho^2) \right\}$

$\mathbf{B}_{609} = \left\{ (\underline{3}, 1), (\underline{2}, \sigma), (\underline{2}, \rho), (\underline{2}, \sigma\rho), (\underline{2}, \rho^2), (\underline{3}, \rho^2), (\underline{0}, \sigma\rho^2), (\underline{2}, \sigma\rho^2) \right\}$

$\mathbf{B}_{610} = \left\{ (\underline{3}, 1), (\underline{2}, \sigma), (\underline{3}, \rho), (\underline{0}, \sigma\rho), (\underline{0}, \rho^2), (\underline{1}, \rho^2), (\underline{3}, \rho^2), (\underline{2}, \sigma\rho^2) \right\}$

$\mathbf{B}_{611} = \left\{ (\underline{3}, 1), (\underline{2}, \sigma), (\underline{3}, \rho), (\underline{0}, \sigma\rho), (\underline{2}, \rho^2), (\underline{0}, \sigma\rho^2), (\underline{1}, \sigma\rho^2), (\underline{3}, \sigma\rho^2) \right\}$

$\mathbf{B}_{612} = \left\{ (\underline{3}, 1), (\underline{2}, \sigma), (\underline{1}, \sigma\rho), (\underline{3}, \sigma\rho), (\underline{0}, \rho^2), (\underline{3}, \rho^2), (\underline{0}, \sigma\rho^2), (\underline{3}, \sigma\rho^2) \right\}$

$\mathbf{B}_{613} = \left\{ (\underline{3}, 1), (\underline{2}, \sigma), (\underline{1}, \sigma\rho), (\underline{3}, \sigma\rho), (\underline{1}, \rho^2), (\underline{2}, \rho^2), (\underline{1}, \sigma\rho^2), (\underline{2}, \sigma\rho^2) \right\}$

$\mathbf{B}_{614} = \left\{ (\underline{3}, 1), (\underline{3}, \sigma), (\underline{0}, \rho), (\underline{1}, \rho), (\underline{2}, \rho), (\underline{3}, \sigma\rho), (\underline{1}, \rho^2), (\underline{1}, \sigma\rho^2) \right\}$

$\mathbf{B}_{615} = \left\{ (\underline{3}, 1), (\underline{3}, \sigma), (\underline{0}, \rho), (\underline{1}, \rho), (\underline{1}, \sigma\rho), (\underline{2}, \sigma\rho), (\underline{3}, \rho^2), (\underline{0}, \sigma\rho^2) \right\}$

$\mathbf{B}_{616} = \left\{ (\underline{3}, 1), (\underline{3}, \sigma), (\underline{0}, \rho), (\underline{2}, \rho), (\underline{0}, \sigma\rho), (\underline{2}, \sigma\rho), (\underline{0}, \rho^2), (\underline{2}, \sigma\rho^2) \right\}$

$\mathbf{B}_{617} = \left\{ (\underline{3}, 1), (\underline{3}, \sigma), (\underline{0}, \rho), (\underline{3}, \rho), (\underline{0}, \rho^2), (\underline{3}, \rho^2), (\underline{1}, \sigma\rho^2), (\underline{3}, \sigma\rho^2) \right\}$

$\mathbf{B}_{618} = \left\{ (\underline{3}, 1), (\underline{3}, \sigma), (\underline{0}, \rho), (\underline{3}, \rho), (\underline{1}, \rho^2), (\underline{2}, \rho^2), (\underline{0}, \sigma\rho^2), (\underline{2}, \sigma\rho^2) \right\}$

$\mathbf{B}_{619} = \left\{ (\underline{3}, 1), (\underline{3}, \sigma), (\underline{0}, \rho), (\underline{0}, \sigma\rho), (\underline{1}, \sigma\rho), (\underline{3}, \sigma\rho), (\underline{2}, \rho^2), (\underline{3}, \sigma\rho^2) \right\}$

$\mathbf{B}_{620} = \left\{ (\underline{3}, 1), (\underline{3}, \sigma), (\underline{1}, \rho), (\underline{2}, \rho), (\underline{3}, \rho), (\underline{2}, \sigma\rho), (\underline{2}, \rho^2), (\underline{3}, \sigma\rho^2) \right\}$

$\mathbf{B}_{621} = \left\{ (\underline{3}, 1), (\underline{3}, \sigma), (\underline{1}, \rho), (\underline{3}, \rho), (\underline{1}, \sigma\rho), (\underline{3}, \sigma\rho), (\underline{0}, \rho^2), (\underline{2}, \sigma\rho^2) \right\}$

$$\mathbf{B}_{622} = \left\{ (\underline{3}, 1), (\underline{3}, \sigma), (\underline{1}, \rho), (\underline{0}, \sigma\rho), (\underline{0}, \rho^2), (\underline{1}, \rho^2), (\underline{0}, \sigma\rho^2), (\underline{3}, \sigma\rho^2) \right\}$$

$$\mathbf{B}_{623} = \left\{ (\underline{3}, 1), (\underline{3}, \sigma), (\underline{1}, \rho), (\underline{0}, \sigma\rho), (\underline{2}, \rho^2), (\underline{3}, \rho^2), (\underline{1}, \sigma\rho^2), (\underline{2}, \sigma\rho^2) \right\}$$

$$\mathbf{B}_{624} = \left\{ (\underline{3}, 1), (\underline{3}, \sigma), (\underline{2}, \rho), (\underline{3}, \rho), (\underline{0}, \sigma\rho), (\underline{3}, \sigma\rho), (\underline{3}, \rho^2), (\underline{0}, \sigma\rho^2) \right\}$$

$$\mathbf{B}_{625} = \left\{ (\underline{3}, 1), (\underline{3}, \sigma), (\underline{2}, \rho), (\underline{1}, \sigma\rho), (\underline{0}, \rho^2), (\underline{2}, \rho^2), (\underline{0}, \sigma\rho^2), (\underline{1}, \sigma\rho^2) \right\}$$

$$\mathbf{B}_{626} = \left\{ (\underline{3}, 1), (\underline{3}, \sigma), (\underline{2}, \rho), (\underline{1}, \sigma\rho), (\underline{1}, \rho^2), (\underline{3}, \rho^2), (\underline{2}, \sigma\rho^2), (\underline{3}, \sigma\rho^2) \right\}$$

$$\mathbf{B}_{627} = \left\{ (\underline{3}, 1), (\underline{3}, \sigma), (\underline{3}, \rho), (\underline{0}, \sigma\rho), (\underline{1}, \sigma\rho), (\underline{2}, \sigma\rho), (\underline{1}, \rho^2), (\underline{1}, \sigma\rho^2) \right\}$$

$$\mathbf{B}_{628} = \left\{ (\underline{3}, 1), (\underline{3}, \sigma), (\underline{2}, \sigma\rho), (\underline{3}, \sigma\rho), (\underline{0}, \rho^2), (\underline{1}, \rho^2), (\underline{2}, \rho^2), (\underline{3}, \rho^2) \right\}$$

$$\mathbf{B}_{629} = \left\{ (\underline{3}, 1), (\underline{3}, \sigma), (\underline{2}, \sigma\rho), (\underline{3}, \sigma\rho), (\underline{0}, \sigma\rho^2), (\underline{1}, \sigma\rho^2), (\underline{2}, \sigma\rho^2), (\underline{3}, \sigma\rho^2) \right\}$$

$$\mathbf{B}_{630} = \left\{ (\underline{0}, \sigma), (\underline{1}, \sigma), (\underline{2}, \sigma), (\underline{3}, \sigma), (\underline{0}, \rho), (\underline{2}, \rho), (\underline{0}, \sigma\rho), (\underline{3}, \sigma\rho) \right\}$$

$$\mathbf{B}_{631} = \left\{ (\underline{0}, \sigma), (\underline{1}, \sigma), (\underline{2}, \sigma), (\underline{3}, \sigma), (\underline{1}, \rho), (\underline{3}, \rho), (\underline{1}, \sigma\rho), (\underline{2}, \sigma\rho) \right\}$$

$$\mathbf{B}_{632} = \left\{ (\underline{0}, \sigma), (\underline{1}, \sigma), (\underline{2}, \sigma), (\underline{3}, \sigma), (\underline{0}, \rho^2), (\underline{0}, \sigma\rho^2), (\underline{1}, \sigma\rho^2), (\underline{3}, \sigma\rho^2) \right\}$$

$$\mathbf{B}_{633} = \left\{ (\underline{0}, \sigma), (\underline{1}, \sigma), (\underline{2}, \sigma), (\underline{3}, \sigma), (\underline{1}, \rho^2), (\underline{2}, \rho^2), (\underline{3}, \rho^2), (\underline{2}, \sigma\rho^2) \right\}$$

$$\mathbf{B}_{634} = \left\{ (\underline{0}, \sigma), (\underline{1}, \sigma), (\underline{0}, \rho), (\underline{1}, \rho), (\underline{2}, \rho), (\underline{1}, \sigma\rho), (\underline{1}, \rho^2), (\underline{3}, \rho^2) \right\}$$

$$\mathbf{B}_{635} = \left\{ (\underline{0}, \sigma), (\underline{1}, \sigma), (\underline{0}, \rho), (\underline{1}, \rho), (\underline{2}, \sigma\rho), (\underline{3}, \sigma\rho), (\underline{0}, \sigma\rho^2), (\underline{1}, \sigma\rho^2) \right\}$$

$$\mathbf{B}_{636} = \left\{ (\underline{0}, \sigma), (\underline{1}, \sigma), (\underline{0}, \rho), (\underline{2}, \rho), (\underline{3}, \rho), (\underline{2}, \sigma\rho), (\underline{2}, \rho^2), (\underline{2}, \sigma\rho^2) \right\}$$

$$\mathbf{B}_{637} = \left\{ (\underline{0}, \sigma), (\underline{1}, \sigma), (\underline{0}, \rho), (\underline{3}, \rho), (\underline{1}, \sigma\rho), (\underline{3}, \sigma\rho), (\underline{0}, \rho^2), (\underline{3}, \sigma\rho^2) \right\}$$

$$\mathbf{B}_{638} = \left\{ (\underline{0}, \sigma), (\underline{1}, \sigma), (\underline{0}, \rho), (\underline{0}, \sigma\rho), (\underline{0}, \rho^2), (\underline{1}, \rho^2), (\underline{0}, \sigma\rho^2), (\underline{2}, \sigma\rho^2) \right\}$$

$$\mathbf{B}_{639} = \left\{ (\underline{0}, \sigma), (\underline{1}, \sigma), (\underline{0}, \rho), (\underline{0}, \sigma\rho), (\underline{2}, \rho^2), (\underline{3}, \rho^2), (\underline{1}, \sigma\rho^2), (\underline{3}, \sigma\rho^2) \right\}$$

$$\mathbf{B}_{640} = \left\{ (\underline{0}, \sigma), (\underline{1}, \sigma), (\underline{1}, \rho), (\underline{2}, \rho), (\underline{0}, \sigma\rho), (\underline{2}, \sigma\rho), (\underline{0}, \rho^2), (\underline{3}, \sigma\rho^2) \right\}$$

$$\mathbf{B}_{641} = \left\{ (\underline{0}, \sigma), (\underline{1}, \sigma), (\underline{1}, \rho), (\underline{3}, \rho), (\underline{0}, \rho^2), (\underline{3}, \rho^2), (\underline{1}, \sigma\rho^2), (\underline{2}, \sigma\rho^2) \right\}$$

$$\mathbf{B}_{642} = \left\{ (\underline{0}, \sigma), (\underline{1}, \sigma), (\underline{1}, \rho), (\underline{3}, \rho), (\underline{1}, \rho^2), (\underline{2}, \rho^2), (\underline{0}, \sigma\rho^2), (\underline{3}, \sigma\rho^2) \right\}$$

$$\mathbf{B}_{643} = \left\{ (\underline{0}, \sigma), (\underline{1}, \sigma), (\underline{1}, \rho), (\underline{0}, \sigma\rho), (\underline{1}, \sigma\rho), (\underline{3}, \sigma\rho), (\underline{2}, \rho^2), (\underline{2}, \sigma\rho^2) \right\}$$

$$\mathbf{B}_{644} = \left\{ (\underline{0}, \sigma), (\underline{1}, \sigma), (\underline{2}, \rho), (\underline{3}, \rho), (\underline{0}, \sigma\rho), (\underline{1}, \sigma\rho), (\underline{0}, \sigma\rho^2), (\underline{1}, \sigma\rho^2) \right\}$$

$$\mathbf{B}_{645} = \left\{ (\underline{0}, \sigma), (\underline{1}, \sigma), (\underline{2}, \rho), (\underline{3}, \sigma\rho), (\underline{0}, \rho^2), (\underline{2}, \rho^2), (\underline{3}, \rho^2), (\underline{0}, \sigma\rho^2) \right\}$$

$$\mathbf{B}_{646} = \left\{ (\underline{0}, \sigma), (\underline{1}, \sigma), (\underline{2}, \rho), (\underline{3}, \sigma\rho), (\underline{1}, \rho^2), (\underline{1}, \sigma\rho^2), (\underline{2}, \sigma\rho^2), (\underline{3}, \sigma\rho^2) \right\}$$

$$\mathbf{B}_{647} = \left\{ (\underline{0}, \sigma), (\underline{1}, \sigma), (\underline{3}, \rho), (\underline{0}, \sigma\rho), (\underline{2}, \sigma\rho), (\underline{3}, \sigma\rho), (\underline{1}, \rho^2), (\underline{3}, \rho^2) \right\}$$

$$\mathbf{B}_{648} = \left\{ (\underline{0}, \sigma), (\underline{1}, \sigma), (\underline{1}, \sigma\rho), (\underline{2}, \sigma\rho), (\underline{0}, \rho^2), (\underline{1}, \rho^2), (\underline{2}, \rho^2), (\underline{1}, \sigma\rho^2) \right\}$$

$$\mathbf{B}_{649} = \left\{ (\underline{0}, \sigma), (\underline{1}, \sigma), (\underline{1}, \sigma\rho), (\underline{2}, \sigma\rho), (\underline{3}, \rho^2), (\underline{0}, \sigma\rho^2), (\underline{2}, \sigma\rho^2), (\underline{3}, \sigma\rho^2) \right\}$$

$$\mathbf{B}_{650} = \left\{ (\underline{0}, \sigma), (\underline{2}, \sigma), (\underline{0}, \rho), (\underline{1}, \rho), (\underline{3}, \rho), (\underline{3}, \sigma\rho), (\underline{1}, \rho^2), (\underline{2}, \sigma\rho^2) \right\}$$

$$\mathbf{B}_{651} = \left\{ (\underline{0}, \sigma), (\underline{2}, \sigma), (\underline{0}, \rho), (\underline{1}, \rho), (\underline{0}, \sigma\rho), (\underline{1}, \sigma\rho), (\underline{0}, \sigma\rho^2), (\underline{3}, \sigma\rho^2) \right\}$$

$$\mathbf{B}_{652} = \left\{ (\underline{0}, \sigma), (\underline{2}, \sigma), (\underline{0}, \rho), (\underline{2}, \rho), (\underline{0}, \rho^2), (\underline{3}, \rho^2), (\underline{2}, \sigma\rho^2), (\underline{3}, \sigma\rho^2) \right\}$$

$$\mathbf{B}_{653} = \left\{ (\underline{0}, \sigma), (\underline{2}, \sigma), (\underline{0}, \rho), (\underline{2}, \rho), (\underline{1}, \rho^2), (\underline{2}, \rho^2), (\underline{0}, \sigma\rho^2), (\underline{1}, \sigma\rho^2) \right\}$$

$$\mathbf{B}_{654} = \left\{ (\underline{0}, \sigma), (\underline{2}, \sigma), (\underline{0}, \rho), (\underline{3}, \rho), (\underline{0}, \sigma\rho), (\underline{2}, \sigma\rho), (\underline{0}, \rho^2), (\underline{1}, \sigma\rho^2) \right\}$$

$$\mathbf{B}_{655} = \left\{ (\underline{0}, \sigma), (\underline{2}, \sigma), (\underline{0}, \rho), (\underline{1}, \sigma\rho), (\underline{2}, \sigma\rho), (\underline{3}, \sigma\rho), (\underline{2}, \rho^2), (\underline{3}, \rho^2) \right\}$$

$$\mathbf{B}_{656} = \left\{ (\underline{0}, \sigma), (\underline{2}, \sigma), (\underline{1}, \rho), (\underline{2}, \rho), (\underline{3}, \rho), (\underline{0}, \sigma\rho), (\underline{2}, \rho^2), (\underline{3}, \rho^2) \right\}$$

$$\mathbf{B}_{657} = \left\{ (\underline{0}, \sigma), (\underline{2}, \sigma), (\underline{1}, \rho), (\underline{2}, \rho), (\underline{1}, \sigma\rho), (\underline{3}, \sigma\rho), (\underline{0}, \rho^2), (\underline{1}, \sigma\rho^2) \right\}$$

$$\mathbf{B}_{658} = \left\{ (\underline{0}, \sigma), (\underline{2}, \sigma), (\underline{1}, \rho), (\underline{2}, \sigma\rho), (\underline{0}, \rho^2), (\underline{1}, \rho^2), (\underline{3}, \rho^2), (\underline{0}, \sigma\rho^2) \right\}$$

$$\mathbf{B}_{659} = \left\{ (\underline{0}, \sigma), (\underline{2}, \sigma), (\underline{1}, \rho), (\underline{2}, \sigma\rho), (\underline{2}, \rho^2), (\underline{1}, \sigma\rho^2), (\underline{2}, \sigma\rho^2), (\underline{3}, \sigma\rho^2) \right\}$$

$$\mathbf{B}_{660} = \left\{ (\underline{0}, \sigma), (\underline{2}, \sigma), (\underline{2}, \rho), (\underline{3}, \rho), (\underline{2}, \sigma\rho), (\underline{3}, \sigma\rho), (\underline{0}, \sigma\rho^2), (\underline{3}, \sigma\rho^2) \right\}$$

$$\mathbf{B}_{661} = \left\{ (\underline{0}, \sigma), (\underline{2}, \sigma), (\underline{2}, \rho), (\underline{0}, \sigma\rho), (\underline{1}, \sigma\rho), (\underline{2}, \sigma\rho), (\underline{1}, \rho^2), (\underline{2}, \sigma\rho^2) \right\}$$

$$\mathbf{B}_{662} = \left\{ (\underline{0}, \sigma), (\underline{2}, \sigma), (\underline{3}, \rho), (\underline{1}, \sigma\rho), (\underline{0}, \rho^2), (\underline{2}, \rho^2), (\underline{0}, \sigma\rho^2), (\underline{2}, \sigma\rho^2) \right\}$$

$$\mathbf{B}_{663} = \left\{ (\underline{0}, \sigma), (\underline{2}, \sigma), (\underline{3}, \rho), (\underline{1}, \sigma\rho), (\underline{1}, \rho^2), (\underline{3}, \rho^2), (\underline{1}, \sigma\rho^2), (\underline{3}, \sigma\rho^2) \right\}$$

$$\mathbf{B}_{664} = \left\{ (\underline{0}, \sigma), (\underline{2}, \sigma), (\underline{0}, \sigma\rho), (\underline{3}, \sigma\rho), (\underline{0}, \rho^2), (\underline{1}, \rho^2), (\underline{2}, \rho^2), (\underline{3}, \sigma\rho^2) \right\}$$

$$\mathbf{B}_{665} = \left\{ (\underline{0}, \sigma), (\underline{2}, \sigma), (\underline{0}, \sigma\rho), (\underline{3}, \sigma\rho), (\underline{3}, \rho^2), (\underline{0}, \sigma\rho^2), (\underline{1}, \sigma\rho^2), (\underline{2}, \sigma\rho^2) \right\}$$

$$\mathbf{B}_{666} = \left\{ (\underline{0}, \sigma), (\underline{3}, \sigma), (\underline{0}, \rho), (\underline{1}, \rho), (\underline{2}, \rho), (\underline{3}, \rho), (\underline{0}, \rho^2), (\underline{0}, \sigma\rho^2) \right\}$$

$$\mathbf{B}_{667} = \left\{ (\underline{0}, \sigma), (\underline{3}, \sigma), (\underline{0}, \rho), (\underline{1}, \rho), (\underline{0}, \sigma\rho), (\underline{2}, \sigma\rho), (\underline{1}, \rho^2), (\underline{2}, \rho^2) \right\}$$

$$\mathbf{B}_{668} = \left\{ (\underline{0}, \sigma), (\underline{3}, \sigma), (\underline{0}, \rho), (\underline{2}, \rho), (\underline{1}, \sigma\rho), (\underline{2}, \sigma\rho), (\underline{1}, \sigma\rho^2), (\underline{3}, \sigma\rho^2) \right\}$$

$$\mathbf{B}_{669} = \left\{ (\underline{0}, \sigma), (\underline{3}, \sigma), (\underline{0}, \rho), (\underline{3}, \rho), (\underline{0}, \sigma\rho), (\underline{1}, \sigma\rho), (\underline{3}, \rho^2), (\underline{2}, \sigma\rho^2) \right\}$$

$$\mathbf{B}_{670} = \left\{ (\underline{0}, \sigma), (\underline{3}, \sigma), (\underline{0}, \rho), (\underline{3}, \sigma\rho), (\underline{0}, \rho^2), (\underline{2}, \rho^2), (\underline{1}, \sigma\rho^2), (\underline{2}, \sigma\rho^2) \right\}$$

$$\mathbf{B}_{671} = \left\{ (\underline{0}, \sigma), (\underline{3}, \sigma), (\underline{0}, \rho), (\underline{3}, \sigma\rho), (\underline{1}, \rho^2), (\underline{3}, \rho^2), (\underline{0}, \sigma\rho^2), (\underline{3}, \sigma\rho^2) \right\}$$

$$\mathbf{B}_{672} = \left\{ (\underline{0}, \sigma), (\underline{3}, \sigma), (\underline{1}, \rho), (\underline{2}, \rho), (\underline{2}, \sigma\rho), (\underline{3}, \sigma\rho), (\underline{3}, \rho^2), (\underline{2}, \sigma\rho^2) \right\}$$

$$\mathbf{B}_{673} = \left\{ (\underline{0}, \sigma), (\underline{3}, \sigma), (\underline{1}, \rho), (\underline{3}, \rho), (\underline{0}, \sigma\rho), (\underline{3}, \sigma\rho), (\underline{1}, \sigma\rho^2), (\underline{3}, \sigma\rho^2) \right\}$$

$$\mathbf{B}_{674} = \left\{ (\underline{0}, \sigma), (\underline{3}, \sigma), (\underline{1}, \rho), (\underline{1}, \sigma\rho), (\underline{0}, \rho^2), (\underline{2}, \rho^2), (\underline{3}, \rho^2), (\underline{3}, \sigma\rho^2) \right\}$$

$$\mathbf{B}_{675} = \left\{ (\underline{0}, \sigma), (\underline{3}, \sigma), (\underline{1}, \rho), (\underline{1}, \sigma\rho), (\underline{1}, \rho^2), (\underline{0}, \sigma\rho^2), (\underline{1}, \sigma\rho^2), (\underline{2}, \sigma\rho^2) \right\}$$

$$\mathbf{B}_{676} = \left\{ (\underline{0}, \sigma), (\underline{3}, \sigma), (\underline{2}, \rho), (\underline{3}, \rho), (\underline{1}, \sigma\rho), (\underline{3}, \sigma\rho), (\underline{1}, \rho^2), (\underline{2}, \rho^2) \right\}$$

$$\mathbf{B}_{677} = \left\{ (\underline{0}, \sigma), (\underline{3}, \sigma), (\underline{2}, \rho), (\underline{0}, \sigma\rho), (\underline{0}, \rho^2), (\underline{1}, \rho^2), (\underline{3}, \rho^2), (\underline{1}, \sigma\rho^2) \right\}$$

$$\mathbf{B}_{678} = \left\{ (\underline{0}, \sigma), (\underline{3}, \sigma), (\underline{2}, \rho), (\underline{0}, \sigma\rho), (\underline{2}, \rho^2), (\underline{0}, \sigma\rho^2), (\underline{2}, \sigma\rho^2), (\underline{3}, \sigma\rho^2) \right\}$$

$$\mathbf{B}_{679} = \left\{ (\underline{0}, \sigma), (\underline{3}, \sigma), (\underline{3}, \rho), (\underline{2}, \sigma\rho), (\underline{0}, \rho^2), (\underline{1}, \rho^2), (\underline{2}, \sigma\rho^2), (\underline{3}, \sigma\rho^2) \right\}$$

$$\mathbf{B}_{680} = \left\{ (\underline{0}, \sigma), (\underline{3}, \sigma), (\underline{3}, \rho), (\underline{2}, \sigma\rho), (\underline{2}, \rho^2), (\underline{3}, \rho^2), (\underline{0}, \sigma\rho^2), (\underline{1}, \sigma\rho^2) \right\}$$

$$\mathbf{B}_{681} = \left\{ (\underline{0}, \sigma), (\underline{3}, \sigma), (\underline{0}, \sigma\rho), (\underline{1}, \sigma\rho), (\underline{2}, \sigma\rho), (\underline{3}, \sigma\rho), (\underline{0}, \rho^2), (\underline{0}, \sigma\rho^2) \right\}$$

$$\mathbf{B}_{682} = \left\{ (\underline{1}, \sigma), (\underline{2}, \sigma), (\underline{0}, \rho), (\underline{1}, \rho), (\underline{2}, \rho), (\underline{3}, \rho), (\underline{1}, \sigma\rho^2), (\underline{3}, \sigma\rho^2) \right\}$$

$$\mathbf{B}_{683} = \left\{ (\underline{1}, \sigma), (\underline{2}, \sigma), (\underline{0}, \rho), (\underline{1}, \rho), (\underline{0}, \sigma\rho), (\underline{2}, \sigma\rho), (\underline{3}, \rho^2), (\underline{2}, \sigma\rho^2) \right\}$$

$$\mathbf{B}_{684} = \left\{ (\underline{1}, \sigma), (\underline{2}, \sigma), (\underline{0}, \rho), (\underline{2}, \rho), (\underline{1}, \sigma\rho), (\underline{2}, \sigma\rho), (\underline{0}, \rho^2), (\underline{0}, \sigma\rho^2) \right\}$$

$$\mathbf{B}_{685} = \left\{ (\underline{1}, \sigma), (\underline{2}, \sigma), (\underline{0}, \rho), (\underline{3}, \rho), (\underline{0}, \sigma\rho), (\underline{1}, \sigma\rho), (\underline{1}, \rho^2), (\underline{2}, \rho^2) \right\}$$

$$\mathbf{B}_{686} = \left\{ (\underline{1}, \sigma), (\underline{2}, \sigma), (\underline{0}, \rho), (\underline{3}, \sigma\rho), (\underline{0}, \rho^2), (\underline{1}, \rho^2), (\underline{3}, \rho^2), (\underline{1}, \sigma\rho^2) \right\}$$

$$\mathbf{B}_{687} = \left\{ (\underline{1}, \sigma), (\underline{2}, \sigma), (\underline{0}, \rho), (\underline{3}, \sigma\rho), (\underline{2}, \rho^2), (\underline{0}, \sigma\rho^2), (\underline{2}, \sigma\rho^2), (\underline{3}, \sigma\rho^2) \right\}$$

$$\mathbf{B_{688}} = \left\{ (\underline{1}, \sigma), (\underline{2}, \sigma), (\underline{1}, \rho), (\underline{2}, \rho), (\underline{2}, \sigma\rho), (\underline{3}, \sigma\rho), (\underline{1}, \rho^2), (\underline{2}, \rho^2) \right\}$$

$$\mathbf{B_{689}} = \left\{ (\underline{1}, \sigma), (\underline{2}, \sigma), (\underline{1}, \rho), (\underline{3}, \rho), (\underline{0}, \sigma\rho), (\underline{3}, \sigma\rho), (\underline{0}, \rho^2), (\underline{0}, \sigma\rho^2) \right\}$$

$$\mathbf{B_{690}} = \left\{ (\underline{1}, \sigma), (\underline{2}, \sigma), (\underline{1}, \rho), (\underline{1}, \sigma\rho), (\underline{0}, \rho^2), (\underline{1}, \rho^2), (\underline{2}, \sigma\rho^2), (\underline{3}, \sigma\rho^2) \right\}$$

$$\mathbf{B_{691}} = \left\{ (\underline{1}, \sigma), (\underline{2}, \sigma), (\underline{1}, \rho), (\underline{1}, \sigma\rho), (\underline{2}, \rho^2), (\underline{3}, \rho^2), (\underline{0}, \sigma\rho^2), (\underline{1}, \sigma\rho^2) \right\}$$

$$\mathbf{B_{692}} = \left\{ (\underline{1}, \sigma), (\underline{2}, \sigma), (\underline{2}, \rho), (\underline{3}, \rho), (\underline{1}, \sigma\rho), (\underline{3}, \sigma\rho), (\underline{3}, \rho^2), (\underline{2}, \sigma\rho^2) \right\}$$

$$\mathbf{B_{693}} = \left\{ (\underline{1}, \sigma), (\underline{2}, \sigma), (\underline{2}, \rho), (\underline{0}, \sigma\rho), (\underline{0}, \rho^2), (\underline{2}, \rho^2), (\underline{1}, \sigma\rho^2), (\underline{2}, \sigma\rho^2) \right\}$$

$$\mathbf{B_{694}} = \left\{ (\underline{1}, \sigma), (\underline{2}, \sigma), (\underline{2}, \rho), (\underline{0}, \sigma\rho), (\underline{1}, \rho^2), (\underline{3}, \rho^2), (\underline{0}, \sigma\rho^2), (\underline{3}, \sigma\rho^2) \right\}$$

$$\mathbf{B_{695}} = \left\{ (\underline{1}, \sigma), (\underline{2}, \sigma), (\underline{3}, \rho), (\underline{2}, \sigma\rho), (\underline{0}, \rho^2), (\underline{2}, \rho^2), (\underline{3}, \rho^2), (\underline{3}, \sigma\rho^2) \right\}$$

$$\mathbf{B_{696}} = \left\{ (\underline{1}, \sigma), (\underline{2}, \sigma), (\underline{3}, \rho), (\underline{2}, \sigma\rho), (\underline{1}, \rho^2), (\underline{0}, \sigma\rho^2), (\underline{1}, \sigma\rho^2), (\underline{2}, \sigma\rho^2) \right\}$$

$$\mathbf{B_{697}} = \left\{ (\underline{1}, \sigma), (\underline{2}, \sigma), (\underline{0}, \sigma\rho), (\underline{1}, \sigma\rho), (\underline{2}, \sigma\rho), (\underline{3}, \sigma\rho), (\underline{1}, \sigma\rho^2), (\underline{3}, \sigma\rho^2) \right\}$$

$$\mathbf{B_{698}} = \left\{ (\underline{1}, \sigma), (\underline{3}, \sigma), (\underline{0}, \rho), (\underline{1}, \rho), (\underline{3}, \rho), (\underline{3}, \sigma\rho), (\underline{2}, \rho^2), (\underline{3}, \rho^2) \right\}$$

$$\mathbf{B_{699}} = \left\{ (\underline{1}, \sigma), (\underline{3}, \sigma), (\underline{0}, \rho), (\underline{1}, \rho), (\underline{0}, \sigma\rho), (\underline{1}, \sigma\rho), (\underline{0}, \rho^2), (\underline{1}, \sigma\rho^2) \right\}$$

$$\mathbf{B_{700}} = \left\{ (\underline{1}, \sigma), (\underline{3}, \sigma), (\underline{0}, \rho), (\underline{2}, \rho), (\underline{0}, \rho^2), (\underline{1}, \rho^2), (\underline{2}, \rho^2), (\underline{3}, \sigma\rho^2) \right\}$$

$$\mathbf{B_{701}} = \left\{ (\underline{1}, \sigma), (\underline{3}, \sigma), (\underline{0}, \rho), (\underline{2}, \rho), (\underline{3}, \rho^2), (\underline{0}, \sigma\rho^2), (\underline{1}, \sigma\rho^2), (\underline{2}, \sigma\rho^2) \right\}$$

$$\mathbf{B_{702}} = \left\{ (\underline{1}, \sigma), (\underline{3}, \sigma), (\underline{0}, \rho), (\underline{3}, \rho), (\underline{0}, \sigma\rho), (\underline{2}, \sigma\rho), (\underline{0}, \sigma\rho^2), (\underline{3}, \sigma\rho^2) \right\}$$

$$\mathbf{B_{703}} = \left\{ (\underline{1}, \sigma), (\underline{3}, \sigma), (\underline{0}, \rho), (\underline{1}, \sigma\rho), (\underline{2}, \sigma\rho), (\underline{3}, \sigma\rho), (\underline{1}, \rho^2), (\underline{2}, \sigma\rho^2) \right\}$$

$$\mathbf{B_{704}} = \left\{ (\underline{1}, \sigma), (\underline{3}, \sigma), (\underline{1}, \rho), (\underline{2}, \rho), (\underline{3}, \rho), (\underline{0}, \sigma\rho), (\underline{1}, \rho^2), (\underline{2}, \sigma\rho^2) \right\}$$

$$\mathbf{B_{705}} = \left\{ (\underline{1}, \sigma), (\underline{3}, \sigma), (\underline{1}, \rho), (\underline{2}, \rho), (\underline{1}, \sigma\rho), (\underline{3}, \sigma\rho), (\underline{0}, \sigma\rho^2), (\underline{3}, \sigma\rho^2) \right\}$$

$$\mathbf{B_{706}} = \left\{ (\underline{1}, \sigma), (\underline{3}, \sigma), (\underline{1}, \rho), (\underline{2}, \sigma\rho), (\underline{0}, \rho^2), (\underline{2}, \rho^2), (\underline{0}, \sigma\rho^2), (\underline{2}, \sigma\rho^2) \right\}$$

$$\mathbf{B_{707}} = \left\{ (\underline{1}, \sigma), (\underline{3}, \sigma), (\underline{1}, \rho), (\underline{2}, \sigma\rho), (\underline{1}, \rho^2), (\underline{3}, \rho^2), (\underline{1}, \sigma\rho^2), (\underline{3}, \sigma\rho^2) \right\}$$

$$\mathbf{B_{708}} = \left\{ (\underline{1}, \sigma), (\underline{3}, \sigma), (\underline{2}, \rho), (\underline{3}, \rho), (\underline{2}, \sigma\rho), (\underline{3}, \sigma\rho), (\underline{0}, \rho^2), (\underline{1}, \sigma\rho^2) \right\}$$

$$\mathbf{B_{709}} = \left\{ (\underline{1}, \sigma), (\underline{3}, \sigma), (\underline{2}, \rho), (\underline{0}, \sigma\rho), (\underline{1}, \sigma\rho), (\underline{2}, \sigma\rho), (\underline{2}, \rho^2), (\underline{3}, \rho^2) \right\}$$

$$\mathbf{B}_{710} = \left\{ (\underline{1}, \sigma), (\underline{3}, \sigma), (\underline{3}, \rho), (\underline{1}, \sigma\rho), (\underline{0}, \rho^2), (\underline{1}, \rho^2), (\underline{3}, \rho^2), (\underline{0}, \sigma\rho^2) \right\}$$

$$\mathbf{B}_{711} = \left\{ (\underline{1}, \sigma), (\underline{3}, \sigma), (\underline{3}, \rho), (\underline{1}, \sigma\rho), (\underline{2}, \rho^2), (\underline{1}, \sigma\rho^2), (\underline{2}, \sigma\rho^2), (\underline{3}, \sigma\rho^2) \right\}$$

$$\mathbf{B}_{712} = \left\{ (\underline{1}, \sigma), (\underline{3}, \sigma), (\underline{0}, \sigma\rho), (\underline{3}, \sigma\rho), (\underline{0}, \rho^2), (\underline{3}, \rho^2), (\underline{2}, \sigma\rho^2), (\underline{3}, \sigma\rho^2) \right\}$$

$$\mathbf{B}_{713} = \left\{ (\underline{1}, \sigma), (\underline{3}, \sigma), (\underline{0}, \sigma\rho), (\underline{3}, \sigma\rho), (\underline{1}, \rho^2), (\underline{2}, \rho^2), (\underline{0}, \sigma\rho^2), (\underline{1}, \sigma\rho^2) \right\}$$

$$\mathbf{B}_{714} = \left\{ (\underline{2}, \sigma), (\underline{3}, \sigma), (\underline{0}, \rho), (\underline{1}, \rho), (\underline{2}, \rho), (\underline{1}, \sigma\rho), (\underline{2}, \rho^2), (\underline{2}, \sigma\rho^2) \right\}$$

$$\mathbf{B}_{715} = \left\{ (\underline{2}, \sigma), (\underline{3}, \sigma), (\underline{0}, \rho), (\underline{1}, \rho), (\underline{2}, \sigma\rho), (\underline{3}, \sigma\rho), (\underline{0}, \rho^2), (\underline{3}, \sigma\rho^2) \right\}$$

$$\mathbf{B}_{716} = \left\{ (\underline{2}, \sigma), (\underline{3}, \sigma), (\underline{0}, \rho), (\underline{2}, \rho), (\underline{3}, \rho), (\underline{2}, \sigma\rho), (\underline{1}, \rho^2), (\underline{3}, \rho^2) \right\}$$

$$\mathbf{B}_{717} = \left\{ (\underline{2}, \sigma), (\underline{3}, \sigma), (\underline{0}, \rho), (\underline{3}, \rho), (\underline{1}, \sigma\rho), (\underline{3}, \sigma\rho), (\underline{0}, \sigma\rho^2), (\underline{1}, \sigma\rho^2) \right\}$$

$$\mathbf{B}_{718} = \left\{ (\underline{2}, \sigma), (\underline{3}, \sigma), (\underline{0}, \rho), (\underline{0}, \sigma\rho), (\underline{0}, \rho^2), (\underline{2}, \rho^2), (\underline{3}, \rho^2), (\underline{0}, \sigma\rho^2) \right\}$$

$$\mathbf{B}_{719} = \left\{ (\underline{2}, \sigma), (\underline{3}, \sigma), (\underline{0}, \rho), (\underline{0}, \sigma\rho), (\underline{1}, \rho^2), (\underline{1}, \sigma\rho^2), (\underline{2}, \sigma\rho^2), (\underline{3}, \sigma\rho^2) \right\}$$

$$\mathbf{B}_{720} = \left\{ (\underline{2}, \sigma), (\underline{3}, \sigma), (\underline{1}, \rho), (\underline{2}, \rho), (\underline{0}, \sigma\rho), (\underline{2}, \sigma\rho), (\underline{0}, \sigma\rho^2), (\underline{1}, \sigma\rho^2) \right\}$$

$$\mathbf{B}_{721} = \left\{ (\underline{2}, \sigma), (\underline{3}, \sigma), (\underline{1}, \rho), (\underline{3}, \rho), (\underline{0}, \rho^2), (\underline{1}, \rho^2), (\underline{2}, \rho^2), (\underline{1}, \sigma\rho^2) \right\}$$

$$\mathbf{B}_{722} = \left\{ (\underline{2}, \sigma), (\underline{3}, \sigma), (\underline{1}, \rho), (\underline{3}, \rho), (\underline{3}, \rho^2), (\underline{0}, \sigma\rho^2), (\underline{2}, \sigma\rho^2), (\underline{3}, \sigma\rho^2) \right\}$$

$$\mathbf{B}_{723} = \left\{ (\underline{2}, \sigma), (\underline{3}, \sigma), (\underline{1}, \rho), (\underline{0}, \sigma\rho), (\underline{1}, \sigma\rho), (\underline{3}, \sigma\rho), (\underline{1}, \rho^2), (\underline{3}, \rho^2) \right\}$$

$$\mathbf{B}_{724} = \left\{ (\underline{2}, \sigma), (\underline{3}, \sigma), (\underline{2}, \rho), (\underline{3}, \rho), (\underline{0}, \sigma\rho), (\underline{1}, \sigma\rho), (\underline{0}, \rho^2), (\underline{3}, \sigma\rho^2) \right\}$$

$$\mathbf{B}_{725} = \left\{ (\underline{2}, \sigma), (\underline{3}, \sigma), (\underline{2}, \rho), (\underline{3}, \sigma\rho), (\underline{0}, \rho^2), (\underline{1}, \rho^2), (\underline{0}, \sigma\rho^2), (\underline{2}, \sigma\rho^2) \right\}$$

$$\mathbf{B}_{726} = \left\{ (\underline{2}, \sigma), (\underline{3}, \sigma), (\underline{2}, \rho), (\underline{3}, \sigma\rho), (\underline{2}, \rho^2), (\underline{3}, \rho^2), (\underline{1}, \sigma\rho^2), (\underline{3}, \sigma\rho^2) \right\}$$

$$\mathbf{B}_{727} = \left\{ (\underline{2}, \sigma), (\underline{3}, \sigma), (\underline{3}, \rho), (\underline{0}, \sigma\rho), (\underline{2}, \sigma\rho), (\underline{3}, \sigma\rho), (\underline{2}, \rho^2), (\underline{2}, \sigma\rho^2) \right\}$$

$$\mathbf{B}_{728} = \left\{ (\underline{2}, \sigma), (\underline{3}, \sigma), (\underline{1}, \sigma\rho), (\underline{2}, \sigma\rho), (\underline{0}, \rho^2), (\underline{3}, \rho^2), (\underline{1}, \sigma\rho^2), (\underline{2}, \sigma\rho^2) \right\}$$

$$\mathbf{B}_{729} = \left\{ (\underline{2}, \sigma), (\underline{3}, \sigma), (\underline{1}, \sigma\rho), (\underline{2}, \sigma\rho), (\underline{1}, \rho^2), (\underline{2}, \rho^2), (\underline{0}, \sigma\rho^2), (\underline{3}, \sigma\rho^2) \right\}$$

$$\mathbf{B}_{730} = \left\{ (\underline{0}, \rho), (\underline{1}, \rho), (\underline{2}, \rho), (\underline{3}, \rho), (\underline{0}, \sigma\rho), (\underline{1}, \sigma\rho), (\underline{2}, \sigma\rho), (\underline{3}, \sigma\rho) \right\}$$

$$\mathbf{B}_{731} = \left\{ (\underline{0}, \rho), (\underline{1}, \rho), (\underline{2}, \rho), (\underline{2}, \sigma\rho), (\underline{0}, \rho^2), (\underline{2}, \rho^2), (\underline{3}, \rho^2), (\underline{1}, \sigma\rho^2) \right\}$$

$$\mathbf{B}_{732} = \left\{ (\underline{0}, \rho), (\underline{1}, \rho), (\underline{2}, \rho), (\underline{2}, \sigma\rho), (\underline{1}, \rho^2), (\underline{0}, \sigma\rho^2), (\underline{2}, \sigma\rho^2), (\underline{3}, \sigma\rho^2) \right\}$$

$$\mathbf{B}_{733} = \left\{ (\underline{0}, \rho), (\underline{1}, \rho), (\underline{3}, \rho), (\underline{0}, \sigma\rho), (\underline{0}, \rho^2), (\underline{2}, \rho^2), (\underline{2}, \sigma\rho^2), (\underline{3}, \sigma\rho^2) \right\}$$

$$\mathbf{B}_{734} = \left\{ (\underline{0}, \rho), (\underline{1}, \rho), (\underline{3}, \rho), (\underline{0}, \sigma\rho), (\underline{1}, \rho^2), (\underline{3}, \rho^2), (\underline{0}, \sigma\rho^2), (\underline{1}, \sigma\rho^2) \right\}$$

$$\mathbf{B}_{735} = \left\{ (\underline{0}, \rho), (\underline{1}, \rho), (\underline{1}, \sigma\rho), (\underline{3}, \sigma\rho), (\underline{0}, \rho^2), (\underline{1}, \rho^2), (\underline{2}, \rho^2), (\underline{0}, \sigma\rho^2) \right\}$$

$$\mathbf{B}_{736} = \left\{ (\underline{0}, \rho), (\underline{1}, \rho), (\underline{1}, \sigma\rho), (\underline{3}, \sigma\rho), (\underline{3}, \rho^2), (\underline{1}, \sigma\rho^2), (\underline{2}, \sigma\rho^2), (\underline{3}, \sigma\rho^2) \right\}$$

$$\mathbf{B}_{737} = \left\{ (\underline{0}, \rho), (\underline{2}, \rho), (\underline{3}, \rho), (\underline{1}, \sigma\rho), (\underline{0}, \rho^2), (\underline{1}, \rho^2), (\underline{1}, \sigma\rho^2), (\underline{2}, \sigma\rho^2) \right\}$$

$$\mathbf{B}_{738} = \left\{ (\underline{0}, \rho), (\underline{2}, \rho), (\underline{3}, \rho), (\underline{1}, \sigma\rho), (\underline{2}, \rho^2), (\underline{3}, \rho^2), (\underline{0}, \sigma\rho^2), (\underline{3}, \sigma\rho^2) \right\}$$

$$\mathbf{B}_{739} = \left\{ (\underline{0}, \rho), (\underline{2}, \rho), (\underline{0}, \sigma\rho), (\underline{3}, \sigma\rho), (\underline{0}, \rho^2), (\underline{0}, \sigma\rho^2), (\underline{1}, \sigma\rho^2), (\underline{3}, \sigma\rho^2) \right\}$$

$$\mathbf{B}_{740} = \left\{ (\underline{0}, \rho), (\underline{2}, \rho), (\underline{0}, \sigma\rho), (\underline{3}, \sigma\rho), (\underline{1}, \rho^2), (\underline{2}, \rho^2), (\underline{3}, \rho^2), (\underline{2}, \sigma\rho^2) \right\}$$

$$\mathbf{B}_{741} = \left\{ (\underline{0}, \rho), (\underline{3}, \rho), (\underline{2}, \sigma\rho), (\underline{3}, \sigma\rho), (\underline{0}, \rho^2), (\underline{3}, \rho^2), (\underline{0}, \sigma\rho^2), (\underline{2}, \sigma\rho^2) \right\}$$

$$\mathbf{B}_{742} = \left\{ (\underline{0}, \rho), (\underline{3}, \rho), (\underline{2}, \sigma\rho), (\underline{3}, \sigma\rho), (\underline{1}, \rho^2), (\underline{2}, \rho^2), (\underline{1}, \sigma\rho^2), (\underline{3}, \sigma\rho^2) \right\}$$

$$\mathbf{B}_{743} = \left\{ (\underline{0}, \rho), (\underline{0}, \sigma\rho), (\underline{1}, \sigma\rho), (\underline{2}, \sigma\rho), (\underline{0}, \rho^2), (\underline{1}, \rho^2), (\underline{3}, \rho^2), (\underline{3}, \sigma\rho^2) \right\}$$

$$\mathbf{B}_{744} = \left\{ (\underline{0}, \rho), (\underline{0}, \sigma\rho), (\underline{1}, \sigma\rho), (\underline{2}, \sigma\rho), (\underline{2}, \rho^2), (\underline{0}, \sigma\rho^2), (\underline{1}, \sigma\rho^2), (\underline{2}, \sigma\rho^2) \right\}$$

$$\mathbf{B}_{745} = \left\{ (\underline{1}, \rho), (\underline{2}, \rho), (\underline{3}, \rho), (\underline{3}, \sigma\rho), (\underline{0}, \rho^2), (\underline{1}, \rho^2), (\underline{3}, \rho^2), (\underline{3}, \sigma\rho^2) \right\}$$

$$\mathbf{B}_{746} = \left\{ (\underline{1}, \rho), (\underline{2}, \rho), (\underline{3}, \rho), (\underline{3}, \sigma\rho), (\underline{2}, \rho^2), (\underline{0}, \sigma\rho^2), (\underline{1}, \sigma\rho^2), (\underline{2}, \sigma\rho^2) \right\}$$

$$\mathbf{B}_{747} = \left\{ (\underline{1}, \rho), (\underline{2}, \rho), (\underline{0}, \sigma\rho), (\underline{1}, \sigma\rho), (\underline{0}, \rho^2), (\underline{3}, \rho^2), (\underline{0}, \sigma\rho^2), (\underline{2}, \sigma\rho^2) \right\}$$

$$\mathbf{B}_{748} = \left\{ (\underline{1}, \rho), (\underline{2}, \rho), (\underline{0}, \sigma\rho), (\underline{1}, \sigma\rho), (\underline{1}, \rho^2), (\underline{2}, \rho^2), (\underline{1}, \sigma\rho^2), (\underline{3}, \sigma\rho^2) \right\}$$

$$\mathbf{B}_{749} = \left\{ (\underline{1}, \rho), (\underline{3}, \rho), (\underline{1}, \sigma\rho), (\underline{2}, \sigma\rho), (\underline{0}, \rho^2), (\underline{0}, \sigma\rho^2), (\underline{1}, \sigma\rho^2), (\underline{3}, \sigma\rho^2) \right\}$$

$$\mathbf{B}_{750} = \left\{ (\underline{1}, \rho), (\underline{3}, \rho), (\underline{1}, \sigma\rho), (\underline{2}, \sigma\rho), (\underline{1}, \rho^2), (\underline{2}, \rho^2), (\underline{3}, \rho^2), (\underline{2}, \sigma\rho^2) \right\}$$

$$\mathbf{B}_{751} = \left\{ (\underline{1}, \rho), (\underline{0}, \sigma\rho), (\underline{2}, \sigma\rho), (\underline{3}, \sigma\rho), (\underline{0}, \rho^2), (\underline{1}, \rho^2), (\underline{1}, \sigma\rho^2), (\underline{2}, \sigma\rho^2) \right\}$$

$$\mathbf{B}_{752} = \left\{ (\underline{1}, \rho), (\underline{0}, \sigma\rho), (\underline{2}, \sigma\rho), (\underline{3}, \sigma\rho), (\underline{2}, \rho^2), (\underline{3}, \rho^2), (\underline{0}, \sigma\rho^2), (\underline{3}, \sigma\rho^2) \right\}$$

$$\mathbf{B}_{753} = \left\{ (\underline{2}, \rho), (\underline{3}, \rho), (\underline{0}, \sigma\rho), (\underline{2}, \sigma\rho), (\underline{0}, \rho^2), (\underline{1}, \rho^2), (\underline{2}, \rho^2), (\underline{0}, \sigma\rho^2) \right\}$$

$\mathbf{B}_{754} = \left\{ (\underline{2}, \rho), (\underline{3}, \rho), (\underline{0}, \sigma\rho), (\underline{2}, \sigma\rho), (\underline{3}, \rho^2), (\underline{1}, \sigma\rho^2), (\underline{2}, \sigma\rho^2), (\underline{3}, \sigma\rho^2) \right\}$

$\mathbf{B}_{755} = \left\{ (\underline{2}, \rho), (\underline{1}, \sigma\rho), (\underline{2}, \sigma\rho), (\underline{3}, \sigma\rho), (\underline{0}, \rho^2), (\underline{2}, \rho^2), (\underline{2}, \sigma\rho^2), (\underline{3}, \sigma\rho^2) \right\}$

$\mathbf{B}_{756} = \left\{ (\underline{2}, \rho), (\underline{1}, \sigma\rho), (\underline{2}, \sigma\rho), (\underline{3}, \sigma\rho), (\underline{1}, \rho^2), (\underline{3}, \rho^2), (\underline{0}, \sigma\rho^2), (\underline{1}, \sigma\rho^2) \right\}$

$\mathbf{B}_{757} = \left\{ (\underline{3}, \rho), (\underline{0}, \sigma\rho), (\underline{1}, \sigma\rho), (\underline{3}, \sigma\rho), (\underline{0}, \rho^2), (\underline{2}, \rho^2), (\underline{3}, \rho^2), (\underline{1}, \sigma\rho^2) \right\}$

$\mathbf{B}_{758} = \left\{ (\underline{3}, \rho), (\underline{0}, \sigma\rho), (\underline{1}, \sigma\rho), (\underline{3}, \sigma\rho), (\underline{1}, \rho^2), (\underline{0}, \sigma\rho^2), (\underline{2}, \sigma\rho^2), (\underline{3}, \sigma\rho^2) \right\}$

$\mathbf{B}_{759} = \left\{ (\underline{0}, \rho^2), (\underline{1}, \rho^2), (\underline{2}, \rho^2), (\underline{3}, \rho^2), (\underline{0}, \sigma\rho^2), (\underline{1}, \sigma\rho^2), (\underline{2}, \sigma\rho^2), (\underline{3}, \sigma\rho^2) \right\}$

REFERENCES

WEBSITES

1. **Canadian Mathematical Society** - A New Proof of The Four Colour Theorem by Ashay Dharwadker, Knot No.221, Knot a Braid of Links, 5 October 2000. http://www.cms.math.ca/cgi/kabol/browse.pl?Number=221
2. **The Math Forum** - A New Proof of The Four Colour Theorem by Ashay Dharwadker, Internet Mathematics Library, Group Theory and Graph Theory, 2000. http://mathforum.org/library/view/16622.html
3. **Tölvunot Fréttahorn** - Ný sönnun á setningunni um fjóra liti, Ashay Dharwadker, 2000. http://www.tolvunot.is/adal.Frettahorn.htm
4. **Il Teorema dei Quattro Colori e la Teoria dei Grafi** - An article by Anita Pasotti, 2007. http://www.matematicamente.it/magazine/ottobre2007/Numero04.pdf
5. **Yahoo! Famous Mathematics Problems** - A New Proof of The Four Colour Theorem by Ashay Dharwadker, 2000. http://dir.yahoo.com/Science/Mathematics/Problems__Puzzles__and_Games/Famous_Problems/Four_Color_Theorem/
6. **The Witt Design** - The Steiner system S(5,8,24) explicitly computed by Ashay Dharwadker, 2002. http://www.dharwadker.org/witt.html
7. **Common Systems of Coset Representatives** - Proof of existence of common systems of representatives for the left and right cosets of a finite subgroup of a group by Ashay Dharwadker, 2005. http://www.dharwadker.org/coset.html
8. **Higgs Boson Mass predicted by the Four Color Theorem** - By Ashay Dharwadker and Vladimir Khachatryan, 2009. http://arxiv.org/abs/0912.5189

BOOKS

9. AHLFORS, L.V. *Complex Analysis*, McGraw-Hill Book Company, 1979.
10. DHARWADKER, A. & PIRZADA, S. *Graph Theory*, Orient Longman and Universities Press of India, 2008.
11. LAM, T.Y. *A First Course in Noncommutative Rings*, Springer-Verlag, 1991.
12. ROTMAN, J.J. *An Introduction to the Theory of Groups*, Springer-Verlag, 1995.
13. VAN LINT, J.H. & WILSON, R.M. *A Course in Combinatorics*, Cambridge University Press, 1992.

PAPERS

14. APPEL, K. & HAKEN, W. *Every Planar Map is Four Colorable*, Bull. Amer. Math. Soc. 82(711-712), 1976.
15. DHARWADKER, A. ***Riemann Surfaces***, Electronic Geometry Models, Model 2002.05.001, 2003. http://www.eg-models.de/models/Surfaces/Riemann_Surfaces/2002.05.001/_preview.html
16. DHARWADKER, A. & SMITH, J.D.H. *Split Extensions and Representations of Moufang Loops*, Comm. in Alg. 23(11),4245-4255, 1995.
17. EILENBERG, S. *Extensions of General Algebras*, Ann.Soc.Polon.Math. 21, 1948.
18. EULER, L. *Elementa Doctrinae Solidorum*, Novi comment. acad. sci. Petrop., 4(109-140), 1752.
19. HALL, P. *On Representatives of Subsets*, J. London Math. Soc. 10, 26-30, 1935.
20. RIEMANN, G.F.B. *Grundlagen für eine allgemeine Theorie der Funktionen*

einer veränderlichen complexen Grösse, 1851, Reprinted in Gesammelte Mathematische Werke, Dover, 1953.

21. TITS, J. *Sur les systèmes de Steiner associés aux trois grands groupes de Mathieu*, Rend. Math. e Appl.(5)23, 166-184, 1964.
22. WITT, E. *Die 5-fach transitiven Gruppen von Mathieu*, Abh. Math. Sem. Univ. Hamburg 12, 256-264, 1938.
23. NIEUWOUDT, I. ***On the Maximum Degree Chromatic Number of a Graph***, Ph.D. Thesis, Department of Mathematical Sciences, Stellenbosch University, 2007.
24. DHARWADKER, A. & PIRZADA, S. ***Applications of Graph Theory***, Journal of the Korean Society for Industrial and Applied Mathematics (KSIAM), Vol. 11, No. 4, 2007.

www.ingramcontent.com/pod-product-compliance
Lightning Source LLC
Chambersburg PA
CBHW050738180526
45159CB00003B/1271
* 9 7 8 1 4 6 6 2 6 5 3 0 1 *